The publisher gratefully acknowledges the generous support of the General Endowment Fund of the University of California Press Foundation.

Genesis of the Salk Institute

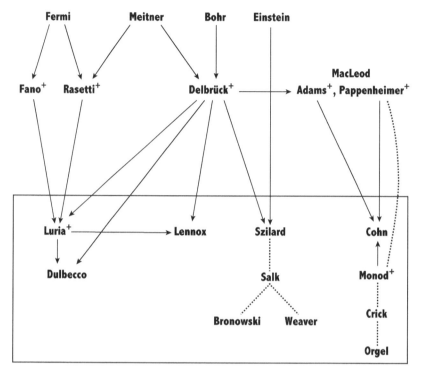

Fermi   Meitner   Bohr   Einstein

MacLeod
Fano[+]   Rasetti[+]   Delbrück[+] ⟶ Adams[+], Pappenheimer[+]

Luria[+] ⟶ Lennox   Szilard   Cohn

Dulbecco   Salk   Monod[+]

Bronowski   Weaver   Crick

Orgel

⟶ Training or collaboration
········· Personal contact
+ Attended the 1946 Cold Spring Harbor Symposium
Inside Box **Became Salk Institute Fellow, resident or non-resident**

Scientific genealogy of the founding team of the Salk Institute

Genesis of the Salk Institute

The Epic of Its Founders

Suzanne Bourgeois

UNIVERSITY OF CALIFORNIA PRESS
Berkeley · Los Angeles · London

University of California Press, one of the most
distinguished university presses in the United States,
enriches lives around the world by advancing scholarship
in the humanities, social sciences, and natural sciences. Its
activities are supported by the UC Press Foundation and
by philanthropic contributions from individuals and
institutions. For more information, visit www.ucpress.edu.

University of California Press
Berkeley and Los Angeles, California

University of California Press, Ltd.
London, England

Library of Congress Cataloging-in-Publication Data

Bourgeois, Suzanne.
    Genesis of the Salk Institute : the epic of its
founders / Suzanne Bourgeois.
        p.   cm.
    Includes bibliographical references and index.
    ISBN 978-0-520-27607-9 (cloth, alk. paper)
    1. Salk Institute for Biological Studies.   2. Biology—
Research—California—San Diego.   3. Research
institutes—California—San Diego.   I. Title.
    QH322.S25B68   2013
    570.72—dc23   2013010766

22 21 20 19 18 17 16 15 14 13
10 9 8 7 6 5 4 3 2 1

To the Salk Institute Founders,
Dreamers of the Greatest Generation

Contents

Illustrations

Foreword

The Salk Institute for Biological Studies is as known and respected in the world of science as it is known and admired in the world of art. This remarkable book by Suzanne Bourgeois takes us through the twenty years following the success of the Salk vaccine against polio and the public adulation it generated.

Jonas Salk was well aware of the need for a new type of institute with a flexible structure that could respond to changing times, as was the polymath Leo Szilard. It might be recalled that it was Szilard who with Einstein had initiated the warning to Roosevelt that led to the development of the atomic bomb. After World War II, Szilard's idea was to bring the emerging field of molecular biology to bear on solutions to the most difficult problems of public health. Jonas was more interested in stressing translational research to deal with the same problems, but with time both ideas were to evolve and broaden. Initially, however, it was this common theme involving human welfare that was to bring them together.

Jonas visited the Institute for Advanced Study in Princeton as a possible model, talked to its president, Robert Oppenheimer, and approached Basil O'Connor, president of the March of Dimes, to explore the idea and possible funding. Then followed offers from the East Coast to the West Coast for the location of his proposed institute, but the one that literally moved Jonas to La Jolla, California, was the donation of a stunning location by the people of San Diego. Meanwhile,

Jonas had approached several colleagues—including Leo Szilard—in the United States, France, and England as potential founding members. A board of trustees was constituted and funding for the building was secured from the March of Dimes through Basil O'Connor. Jonas, after selecting Louis Kahn as the architect, spent years with him planning what was eventually to be the institute as we know and see it today, with its unique and elevating design.

What this short introduction does not reveal, but what this book by Suzanne Bourgeois accomplishes, is to recount the extraordinary story of the ups and downs of each of these moves and the personalities involved where, as Harriet Beecher Stowe put it, "every man had his own quirks and twists," to bring us eventually to the Salk Institute, which is, as pointed out earlier, highly regarded in the worlds of both science and art.

Suzanne Bourgeois, a distinguished biologist, is in a unique position to tell this story. She was at the Pasteur Institute in Paris in 1961 when these enquiries and moves by Jonas Salk were initiated. She personally knew all the characters involved in the odyssey that culminated in the Salk Institute for Biological Studies as we know it today. Personal relationships have been put aside; this book is strictly factual, based on archives, the author's diary, and probing interviews.

Roger Guillemin M.D., Ph.D.
Distinguished Professor, The Salk Institute
Nobel Prize in Physiology or Medicine 1977

Preface and Acknowledgments

This is not a biography of Jonas Salk, nor is it a new version of the story of the polio vaccine. Further, the Salk Institute did not commission this work and it is not a job for hire. This book is a labor of love. It is a personal account of the origins and early days of the Salk Institute for Biological Studies. This remarkable story has never been told, but the fiftieth anniversary of the beginning of the institute's operation—July 1, 1963—makes it timely. This is not a story about the science of the institute; rather, it is about the people.

Many of the characters have passed away, but I knew them all and some were my friends. (I can hardly deny that one of the founders is my husband of fifty years. If that makes my account suspect of bias, so be it.) This account is based on facts that I have painstakingly researched in various archives and in my diary. Archives do not forget or distort the facts; they also do not often tell how and why things happen. Because of my long involvement with the institute and its people, I have been able to collect candid interviews and conduct revealing informal conversations that complement my archival research. The weaving of the events into this story, the choice of facts, and their interpretation are mine. They are based on my experience as a witness and participant in the history of the Salk Institute since 1961, before its inception. That is what puts me in the unique position to narrate this epic.

The historical backdrop for the genesis of the Salk Institute includes World War II, the Cold War, and the Vietnam War. My own upbringing in Belgium during the five years of Nazi occupation makes me a reliable—if biased—witness of the war tragedies that influenced the founders of the institute as well. I was too young to fight but too old to forget.

The idea to write a book about the beginnings of the Salk Institute struck me in 2005, the year of the fiftieth anniversary of the announcement of the Salk polio vaccine in Ann Arbor, Michigan. On that occasion a number of books were published recounting once again Salk's story, which had been so well told for the first time in 1966 by Dick Carter in *Breakthrough: The Saga of Jonas Salk*.[1] Even though Jonas lived for another forty years after the announcement of the vaccine in 1955, all of the books narrating the story of the Salk vaccine said little or nothing about the Salk Institute. It was imperative that the second saga of Jonas Salk be told before it was lost and all the witnesses were gone. Having closed my laboratory by then, I decided to make this my mission.

However, I needed the support and approval of a number of people directly involved with this project, first of all the Salk family. Peter Salk gave me his confidence; he generously granted me interviews, opened to me the institute-related archives of his father's papers, and gave me permission to quote letters and notes. Richard Murphy, the president of the institute at the time, trusted me as well and allowed me access to relevant administrative files and photographic materials. My first problem was to locate administrative archives for the period from 1960 to the mid-1970s. Here I have to acknowledge the enormous help of the institute's Facilities Services team under the leadership of Garry Van Gerpen. The team members are too numerous to mention them all, but I came to greatly appreciate the Bills, the Steves, Stan, Dana, and many others. They helped me to search the dark corners of storage areas where we discovered a series of rusty filing cabinets that contained administrative archives from the early days.

Other teams at the Salk Institute helped me as well throughout the research and writing of this book. From the Information Technology Department I received constant help from Jason Guardiano, who patiently coached me in learning new computer tricks and helped me prepare illustrations; Liz Grabowski in Multimedia Resources performed miracles researching and restoring Salk Institute photographic archives; and Carol Bodas and Rosa Lopez in the Salk Institute Library did detective work to track down articles and books from the past. I am grateful also to the numerous faculty colleagues, alumni, and staff members who granted me interviews or participated in illuminating conversations and gave me unfailing moral support for years.

I also received much help outside the Salk Institute, especially at the library of the University of California, San Diego, where Lynda Claassen and Steve Coy guided me in researching the Mandeville Special Collections. David Rose, the archivist of the March of Dimes, was remarkably generous and supportive throughout my project, and Daniel Demellier always made me feel like my research at the Pasteur Institute Archives was a homecoming.

A turning point in this project was the opportunity given to me by Judith Hodges to establish a partnership between the La Jolla Historical Society and the Salk Institute. This involved the cooperation of John H. Bolthouse, the executive director of the society at the time, and Donald G. Yeckel, a Society Board member and trustee of the Ray Thomas Edwards Foundation. Thanks to a grant from the Edwards Foundation, I was able to produce a number of recordings and transcripts of interviews that were key to reconstructing the early history of the Salk Institute. The Edwards Foundation also provided support for publishing and editorial assistance crucial to the production of my book.

Finally, thank you, Mel, for including me in your dreams.

The Characters

1965—Salvador Luria (1912–1991)
1965—Jerome Wiesner (1915–1994)
1966—Stephen Kuffler (1913–1980)
1968—Daniel Lehrman (1919–1972)

The First Junior Members (1969)

Ursula Bellugi
Suzanne Bourgeois
Walter Eckhart
Stephen Heinemann
David Schubert

The Presidents

Jonas Salk (1960–1962 and 1963–1965)
Gerard Piel (1962–1963)
Augustus Kinzel (1965–1967)
Jerome S. Hardy (pro tem 1967–1968)
Joseph Slater (1968–1972)
Frederic de Hoffmann (1972–1988) (chancellor 1970–1972)

Other Characters Related to the Salk Institute

Donna Salk, Salk's first wife
Peter, Darrell, and Jonathan, Salk's three sons
Françoise Gilot-Salk, Salk's second wife
Lorraine Friedman, Salk's secretary
John Hunt, vice-president 1969–1970
Chantal Hunt, John's wife and friend of Françoise Gilot-Salk
Louis Kahn, architect

NATIONAL FOUNDATION FOR INFANTILE PARALYSIS
(NFIP)

Franklin D. Roosevelt (1882–1945), Founder of the NFIP

(Daniel) Basil O'Connor (1892–1972), First President of the NFIP

Thomas Rivers (1888–1962), Chairman, Committee on Scientific Research

Harry M. Weaver, Research Director

UNIVERSITY OF PITTSBURGH

Edward Harold Litchfield, Chancellor

Frank J. Dixon Jr., Chairman, Department of Pathology

Thaddeus S. Danowski, Professor of Medicine

William S. McEllroy, Dean of the School of Medicine

INSTITUTE FOR ADVANCED STUDY AT PRINCETON

Abraham Flexner (1866–1959), Founding Director, 1930–1939

Albert Einstein (1879–1955), First Permanent Faculty Member

(Julius) Robert Oppenheimer (1904–1967), Director, 1947–1966

Chronology

1955

April 12	Ann Arbor polio vaccine announcement

1956

July 1	Litchfield takes office as chancellor of University of Pittsburgh (Pitt)
Oct.	Salk meets Szilard
Nov.–Dec.	Salk discusses an institute with Litchfield

1957

Jan. 14	Szilard–Doering memo sent to Salk
Feb. 8	Salk declines participation in Szilard's institute plan
May	Salk plans for an institute at Pitt that includes Frank Dixon
Aug.	Salk Hall proposed to house Salk's institute at Pitt
Dec.	Litchfield sets up an advisory committee

1958

Jan.	Salk visits Oppenheimer in Princeton
Feb. 22	Advisory committee meeting
Fall	Negotiations with Pitt fail
Sept.	Salk reminds O'Connor of Szilard's 1957 proposal for an institute

1959

Feb.	Salk visits Oppenheimer and considers an institute outside Pitt
May	Szilard writes to Salk about Revelle and La Jolla
July	O'Connor introduces Salk to Stanford president Sterling
Early Aug.	Salk visits Stanford, meets Sterling, tours Bay Area with Cohn
Aug. 17–21	First visit of Salk to La Jolla to meet the Revelles
Oct.	Salk contacts potential institute Members
Nov. 14–16	Salk's second visit to La Jolla with the O'Connors
Dec.	Salk visits Kahn in Philadelphia

1960

Jan. 2–11	Salk's third visit to La Jolla, meets San Diego city officials
Feb.	Salk's fourth visit to La Jolla with Kahn and O'Connor
March	Salk–Revelle conflict; San Diego City Council hearings
	NF-MOD pledges funds for institute operations and endowment
April	Salk visits Warren Weaver
May 24	Agreement on site among City, Salk, and UCSD
June 7	Municipal ballot vote
July	Salk in London, visits Crick, Bronowski, and Snow

Dec.   Institute for Biology at Torrey Pines; incorporation and bylaws

1961

Jan.   Name change to Institute for Biology at San Diego

April   Salk considers abandoning institute plan

      O'Connor offers to raise funds for the building through the NF-MOD

July   Architectural contract between Kahn and Institute for Biology at San Diego

Aug.–Oct. Lennox and Cohn arrive at the Pasteur Institute in Paris

Oct. 6   Salk visits Fellows at the Pasteur Institute

Oct. 22   Salk meets informally with Fellows in Paris: institute structure defined

Dec. 1   Name change to the Salk Institute for Biological Studies;

      Seven new trustees elected; Weaver designated chairman of the board

Dec. 14   The NF-MOD trustees authorize fifteen-million-dollar fund drive for institute building

Dec. 19   Deed of land granted to the Salk Institute

1962

Feb. 4   First officially recorded Fellows meeting (in Paris): planning and bylaws

Feb. 28   Ground breaking

March 7   Trustees approve amended bylaws and appointment invitations

      Gerard Piel temporarily takes over the institute presidency

April   Contract for laboratory building signed with contractor

May   Major change in Kahn building plans

May 19   Fellows meeting in New York: first institute faculty established

May 31	Press conference in New York at the invitation of Piel
June 1	Institute faculty announcement in *Science*
June 2	Site dedication by Rabbi Morton Cohn
	NF-MOD fund-raising drive for building construction launched and fails
Dec.	Salk moves to La Jolla; excavation and first concrete poured

1963

Jan.	Weaver arrives in La Jolla for a six-week stay
Feb.	Excavation and foundations completed
March 14	NF-MOD pledges eight-million-dollar construction grant; institute to raise seven million dollars
March 30	Fellows turn down offer of space at UCSD in favor of temporary buildings
July 1	Official commencement of operations in barracks
July 22	Barracks near completion
Aug.	Kahn's work on meetinghouse and residences halted
Fall	Lennox and Cohn move to La Jolla from Paris
	Capital campaign to raise seven million dollars in building funds fails

1964

Jan.	Construction loan (ten million dollars) negotiated
	Bronowski moves to La Jolla from London
Feb.	Fellows colloquium: The Nervous System: Facts and Hypotheses
	First Fellows meeting at the institute
May 30	Szilard dies
July	Dulbecco moves to La Jolla from Glasgow
Sept.	Orgel joins the institute

1965

Feb.	Fellows colloquium: Antibody Workshop
July	Construction stopped (fifteen million dollars spent)
Aug.	Kinzel becomes institute president and CEO
Sept.	Luria and Wiesner elected nonresident Fellows
Nov.	Symposium: The Biology of Aging
Dec.	Attempts to attract the Harvard Neurobiology Group fail

1966

June–July	Roman Jakobson's first stay as visiting Fellow
Summer	Move into the north Kahn building completed
Oct.	Kuffler becomes nonresident Fellow

1967

Jan.	Karl Popper's stay as visiting Fellow
Feb.	Fellows colloquium: Studies on Development of Behavior
May	Turbulent times: LSD and art show incidents
June–July	Roman Jakobson's second stay as visiting Fellow
Aug.	Fellows plans completed
Oct.	Kinzel resigns as institute president and trustee

1968

Jan.	Slater appointed president and CEO for a four-year term
Feb.	Fellows meeting, and planning exercise begins
May	Holley appointed resident Fellow, Lehrman nonresident Fellow
	Junior academic ranks created
June	Program of action approved by the trustees

| Sept. | Hunt appointed first vice-president |
| Dec. | McCloy elected trustee |

1969

Jan.	Reproductive physiology meeting
Feb.	Slater moving to New York, desire to preside over the Aspen Institute
	Workshop: Drug Problems and Drug Abuse
March	Hunt elected executive vice-president
April	Council for Biology in Human Affairs (CBHA) planned
July	Meeting: Biological Foundations for Language
Oct.	Slater announces his imminent resignation
	Negotiations to appoint Guillemin approved

1970

Feb.	NF–Salk Institute crisis starts
	CBHA formally established
	John Platt's stay as visiting Fellow
March	Slater to become president of Aspen Institute
	First five junior Members appointed
May	Slater proposes de Hoffmann as chancellor
June	Jonas Salk and Françoise Gilot marry in France
July	Guillemin and his group establish the Neuroendocrinology Lab
Aug.	De Hoffmann appointed chancellor and CEO
	Boardman appointed general secretary of the CBHA

1971

May	NF–Salk Institute crisis continues
July	The Guillemin group moves into south building
Sept.	O'Connor and Nee resign from Salk Institute board

1972

Feb.	Slater resignation accepted and de Hoffmann elected president
March	O'Connor dies in Phoenix
Spring	NF–Salk Institute negotiations resume
	Dulbecco leaves for London
Aug.	Lehrman dies at age fifty-three

Prologue

The Greatest Generation

This generation of Americans has a rendezvous with destiny.
—Franklin Delano Roosevelt[1]

The founders of the Salk Institute are the heroes of this story. They belonged to the generation of the Great Depression and the Second World War. Scientifically, they saw the emergence of the field of molecular biology, and several of our founders played an essential role in what turned out to be a revolution in biology. However, it is World War II—both its horrific events and its aftermath—that marked them most as they became deeply involved in experiences that shaped their view of the future, what they wanted to do, the way they operated, and their capacity for risk. After the war, the commitment of the U.S. government to the veterans in the form of the GI Bill and special training made this generation of Americans the best educated anywhere and at any time in history. These circumstances naturally selected an impressive network of outstanding achievers, and the Salk Institute is part of their legacy.

Who were our founders? What happened to them during World War II? What did they do after the war? This is the story of how and why, born far apart, they eventually gathered in La Jolla in the early 1960s to found an exceptional institution.

Born in the same year, 1914, Jonas Salk[2] and Renato Dulbecco[3] were both medical doctors involved in virus research. Salk was a New Yorker and an alumnus of City College, while Dulbecco was born in Calabria, Italy. The youngest of our founders, Melvin Cohn, was born in New York in 1922.[4] A tough Brooklyn boy, he was, like Salk, a City College alumnus. His friend Edwin Lennox was a southern gentleman born in Savannah, Georgia, in 1920.[5] Both Lennox and Cohn had a background in physics but had turned to immunology by the time they considered joining the Salk Institute. Jacob Bronowski, the oldest of our founders, was born in 1908 in Lodz, Poland; he was a mathematician and a humanist.[6] This group of five individuals represented our first resident faculty, and they called themselves simply "The Fellows."

At the time of its inception, however, the institute had a solid and essential second faculty—nonresident Fellows—who spent considerable time at the institute, were directly involved in its activities, and played a critical role in making the Salk Institute what it was and what it has become. These were senior people who, by the 1960s, were already celebrities. The original group consisted of Warren Weaver, Leo Szilard, Jacques Monod, and Francis Crick. The oldest of our founding non-resident Fellows, Weaver, was born in 1894 in Reedsburg, Wisconsin. Trained as a scientist and mathematician, he was one of the founders of the theory of communication and a science administrator who coined the term "molecular biology" while at the Rockefeller Foundation.[7] Szilard was a Hungarian immigrant born in 1898 in Budapest, a physicist who conceived the atomic bomb before he turned to biology and who was involved in numerous political and scientific issues[8]. Monod and Crick were both from Western Europe. Monod was born in Paris in 1910,[9] while Crick was born in 1916 in a small village close to Northampton, England.[10] Crick, trained as a physicist, gained fame by discovering the structure of the genetic material DNA. Monod, the only founding faculty member formally trained as a biologist, pioneered the field of gene regulation.

World War II began on September 3, 1939, when Britain and France declared war on Germany, two days after Hitler invaded Poland. Fighting in Western Europe started on May 10, 1940, the day Nazi Germany invaded Belgium. I remember, as a little girl living in Brussels, being awakened early that morning by strange loud noises. My mother ran into the room I shared with my sister and yelled, *"C'est la guerre!"*[11] Through the big window in our kitchen I saw the first German airplanes flying low. I heard the first bombs exploding and I felt the ground shaking. For us Belgians, this day would mark the beginning of a five-year nightmare: the occupation, the bombings first by the Luftwaffe and later by the Allies, the fear, the freezing winters, the sleepless nights in the basement, and the scarce food. Soon the Nazis invaded France, while Spain and Italy sided with Hitler, and the war spread to the

Mediterranean basin, North Africa, and eventually Eastern Europe and Russia. Far away Japan attacked islands in the Pacific and finally Pearl Harbor on December 7, 1941, and the United States entered the war.

The invasion of Belgium certainly came as no surprise to Leo Szilard. As early as August 1939, Szilard, then a forty-year-old American physicist, had drafted with his mentor, Albert Einstein, a letter to President Franklin D. Roosevelt warning him of the possibility of using uranium as a powerful source of energy that could lead to the construction of extremely destructive bombs. A few years earlier, in 1933, Szilard had conceived the famous chain reaction that is the principle of the atomic bomb (he patented it one year later). In collaboration with the Italian physicist Enrico Fermi, he invented the nuclear reactor that uses uranium to sustain such a chain reaction and releases atomic energy by nuclear fission. The Einstein-Szilard letter urged the support and development of research in this field and further informed the American president that the richest source of uranium was the Belgian Congo. That Germany had taken over the Czechoslovakian uranium mines and stopped the sale of uranium was also an indication that the Nazis might be involved in nuclear fission research.

The August 2, 1939, letter, conceived by Szilard but signed by Einstein, was to be delivered by Alexander Sachs, an economist and banker and an unofficial advisor and friend of Roosevelt. However, the start of World War II on September 3, four weeks after the letter was written, had delayed Sachs's appointment with the president. It is in that political climate that the letter was eventually delivered in person by Sachs to the Oval Office on October 11. Sachs included a cover letter to the president pointing out the extreme danger of a German invasion of Belgium, an ally that could provide the richest uranium ore in the world. Roosevelt responded promptly and efficiently by immediately setting up the Advisory Committee on Uranium that included Sachs, Szilard, two other influential Hungarian/American physicists, Eugene Wigner, Edward Teller, and representatives of the U.S. Army and Navy. This committee promised U.S. government funding of uranium research, which was the first step in the American program to build the atomic bomb: the Manhattan Project.[12]

The Advisory Committee on Uranium eventually became a section of the Office of Scientific Research and Development (OSRD), a civilian establishment in the Executive Office of the president, headed by Vannevar Bush, who reported directly to Roosevelt and who had become an influential military advisor to the president. By 1941, Bush was convinced that the United States should undertake an aggressive program to build an atomic bomb, and by the end of that year he had obtained Roosevelt's approval. At the request of Bush, the project was turned over to the Army in June 1942. The Army Corps of Engineers appointed to head the project Colonel Leslie R. Groves, a West Point graduate who had masterfully directed the construction of the Pentagon. Quickly promoted to brigadier general, Groves selected J. Robert Oppenheimer as director of research at the Los Alamos Laboratory in New Mexico, established in 1943, where the bomb would be built.

Various aspects of the Manhattan Project were carried out at multiple sites in the United States, including the University of Chicago Metallurgical Laboratory, where Szilard worked. Being led by an army general did not sit well with Szilard, who quickly clashed with Groves. Unable to fire Szilard, Groves tried to get him interned for the duration of the war, but he only succeeded in banning Szilard from entering the Los Alamos Laboratory compound and in getting him to sign over the rights to his nuclear invention patents to the army for one dollar!

One of the Fellows did work on the A-bomb in Los Alamos: Edwin Lennox. In 1942 Lennox had started graduate school at the University of Rochester, where the theoretical physicist Victor Weisskopf was a professor. However, the Physics Department was being drained as several senior members of the faculty had left to work on various war-related projects. By 1943 the department closed down. Eventually Lennox received a letter from Weisskopf inviting him to work at an undisclosed location on an unidentified project. In early 1944 he ended up in Los Alamos, where he was assigned, as a civilian, to work in a group led by Weisskopf in the Theoretical Physics Division headed by Hans Bethe.[13] In July 1945, Lennox was sent down to a site near Alamogordo to help set up equipment where the A-bomb was going to be tested. He witnessed the first nuclear explosion, the Trinity test, on July 16. As Lennox recalls, "It was one of those life-changing experiences."

By the time the A-bomb was ready and successfully tested at the Trinity site, Roosevelt had died, Truman had been sworn in as the thirty-second American president on April 12, 1945, and Germany had surrendered on May 7. With the war in Europe won and Japan growing weaker, Szilard made several unsuccessful attempts to stop the use of the A-bomb. On August 6, a bomb was dropped on Hiroshima. Three days later a second bomb was dropped on Nagasaki and Japan surrendered on August 15. While making unnecessary the U.S. invasion of Japan that had been planned for the fall, these two bombings killed an estimated 250,000 people, mostly civilians, and started the era of fear of nuclear wars that lasts to this day. Szilard tried in many ways and for the rest of his life to speak to the conscience of the scientists.

Before the A-bombs were dropped, the Allies had concentrated forces in the Philippines in preparation for the invasion of Japan. Those Allied forces consisted mostly of American troops and included one of our Fellows, Melvin Cohn.[14] In March 1942, when General Douglas MacArthur had left his headquarters in Corregidor to escape the Japanese invaders, he warned, "I shall return." He came back in the fall of 1944 with heavy troops that liberated the Philippines in the largest campaign of the Pacific war, the Battle of Luzon. It took until March 1945 to clear Manila of all Japanese, and Cohn has powerful memories of entering the city under snipers' fire. At the news of the surrender of Japan most American soldiers were sent home, but Cohn was declared essential and was shipped to Hiroshima to help prepare the army's report on the effect of the A-bomb. What struck him most upon entering the ruins of the Japanese city was that "nothing stood higher than my waist, except for a tall and twisted metal structure." That structure still stands, preserved as a memorial: it is the frame of a tall building, the Genbaku Dome, which withstood the explosion because the blast took place directly above that structure.

Three months after the A-bombs had been dropped, Jacob Bronowski was sent to Nagasaki to report on the effects of the bombings to the British government. England had made important contributions to the development of the A-bomb, both by influencing the initiation of the Manhattan Project and by carrying out its own research program on uranium. The British wanted their own report. As scientific deputy to the British Chiefs of Staff Mission to Japan in November 1945, Bronowski participated in writing the report "The Effects of the Atomic Bombs at Hiroshima and Nagasaki," and he vividly recounted his impressions of Nagasaki.[15] During the war he had also used his expertise as a mathematician to evaluate the most destructive means to bomb German targets on the ground.

Another Fellow mathematician, Warren Weaver, played an active role in protecting London from German bombs. As a statistician, he was one of the first to be appointed to Bush's OSRD to organize and direct a section called the "fire-control section." That group was concerned mostly with the design and test of devices that increase the efficiency of antiaircraft fire. The bombing of England had started in July 1940, causing disastrous damage to the London area. During the Blitz in March 1941, Weaver sailed to England with a group of scientists who, on the night of April 16, witnessed the worst bombing raid on London: from nine P.M. until five A.M. hundreds of German bombers flew over London, leading Weaver to write, "From the roof of our hotel one saw a ring of huge fires." In June 1944 the Germans began to use "buzz bombs," unmanned missiles preset to detonate over London. I remember those well because we could hear them fly over Belgium, where many exploded as well. By that time several devices had been developed by the OSRD that tracked and computed the position of planes and missiles and vastly increased the chance of their interception by the antiaircraft guns installed on the east coast of England.

While Francis Crick was a graduate student in physics at University College London (UCL) in 1939, England had declared war on Germany. Before he had completed his "scientifically boring" research, not only did his laboratory close but the apparatus that was the product of his doctorate research was destroyed by German explosives.[16] The German navy was using U-boats and mines to block merchant ships in an attempt to starve Britain into submission. Crick was appointed as a civilian to the Admiralty Research Laboratory. He was first assigned to the Minesweeping Division, and by 1941 he was attached to an ambitious group of the Mine Design Department located near Havant, a small town close to Portsmouth on the south English coast. Here Crick was able to exercise his ingenuity and logical thinking to successfully design new types of mines that sank more than a thousand German and Italian ships. Time spent in the Mine Design Department allowed Crick to interact with distinguished mathematicians and physicists, and this proved critical to his career as it generated confidence and enthusiasm for his research.

Across the Channel, France had been attacked in the spring of 1940 while Jacques Monod was at the Sorbonne, completing the work on bacterial growth for which he was awarded his Ph.D. in 1941. As France immediately negotiated an armistice with Germany, only the north of France—including Paris—was militarily occupied, leaving most of the country under a French regime that cooperated with Germany. This makes the World War II history in France complex, as resistance was not openly encouraged. However, as early as October 1940 Monod was already involved with a resistance group and was interrogated by the Gestapo, but he managed to escape. All through the war Monod resisted actively with a group from the Sorbonne organized by the Communist Party. After multiple arrests of leaders from that group, it had become dangerous to go back to the Sorbonne, and Monod started to work at the Pasteur Institute in the lab of André Lwoff. Through his brother, a resistance leader who lived in Geneva, Monod was able to furnish precious information to London. He also played a dangerous role in the liberation of Paris when he occupied the war ministry and arrested French leaders who had cooperated with the Nazis.

If the political situation in France during World War II was ambiguous, it was even worse in Italy, where the Fascist Mussolini had been dictator since 1925, when Renato Dulbecco was still a teenager. In June 1940, while Germany appeared victorious, Italy declared war on France. As a recently graduated M.D., Dulbecco was immediately drafted as a physician with an infantry regiment and sent to the French border, where he received his baptism by fire in Italy's disastrous attempt to invade France. However, as expected from its new regime, France negotiated an armistice with Italy within two weeks. By March 1943 Dulbecco was sent to Ukraine with the Italian forces fighting on the side of Germany on the Russian front. At first he regarded that campaign as a great adventure,[17] but he was soon disenchanted as he experienced bitter cold, bombings, and the atrocities of war. By September 1943 Italy surrendered to the Allies and, in an about-face, declared war on Germany! In the total confusion that followed, Dulbecco started helping the anti-Fascist partisans until the Allies liberated Italy from German occupation and Fascism in April 1945.

Meanwhile Jonas Salk was doing what he could do best: developing an anti-influenza vaccine for the American Army fighting in Europe. After his internship at Mount Sinai in 1941, Salk had obtained a temporary job with Thomas Francis, his former mentor from New York University who had moved to the University of Michigan. With the United States at war, Francis was made head of the army's Commission on Influenza, and in spring 1942 he helped Salk obtain an occupational deferment to study the influenza virus. This was, indeed, a field directly relevant to the war effort, as flu had devastated the U.S. Army during the First World War, and Francis was the recognized U.S. expert on the human flu virus. It is while working with Francis on the flu vaccine that Salk received the training he would later need to develop the polio vaccine, a crucial step in the genesis of the Salk Institute.

It is poignant that, by the fortunes of war, Dulbecco was with the German army on the Russian front, whereas Cohn was with MacArthur in the Philippines and Bronowski was seeking the maximum destruction of Germany while Weaver was working on shielding

London. It is ironic that Cohn and Bronowski were involved in reporting on the effect of bombs that Szilard and Lennox took part in building. Meanwhile, Monod and Crick were in harm's way and interrupted their careers to protect France and England from Nazism and destruction. All had horrendous experiences that were to change their lives and transform the world, and the painstaking war-oriented flu vaccine work of Salk was to lead to the successful polio vaccine and to the creation of an institute that would contribute to making our planet a little more livable in the atomic age.

What kind of an institute did Salk have in mind? How did his vision evolve? How did he end up in unheard-of La Jolla? How did he attract such a stellar founding faculty? What did he have to offer? Was he bluffing or could he deliver?

This is the story of the difficult birth and growing pains of the Salk Institute from its origin to its realization and early history, covering a period of about twenty years, starting after the success of the polio vaccine in 1955 until the mid-1970s.

CHAPTER 1

Before and after Ann Arbor

Le génie n'est qu'une grande aptitude à la patience.[1]
—Buffon

Poliomyelitis (polio) in the United States has essentially disappeared. Actually, the neurological form of polio was rare, and most victims of the disease survived.[2] However, polio victims were typically children, and some of the survivors were left with crippling paralysis. Polio could leave shocking and heartbreaking evidence: little kids struggling to walk with braces and crutches or confined to monstrous iron lungs. Understandably, all parents were terrified, and it is that terror that inspired an entirely new approach to fund raising. Millions of ordinary people, not millionaires, united to collect large sums of money one dime at a time. It is that public support, orchestrated by the National Foundation for Infantile Paralysis (NFIP), which led to a solution to the disease and to the creation of the Salk Institute.

The story of the polio vaccine has been told many times.[3] Its drama culminated on April 12, 1955, in Ann Arbor, Michigan, where it was announced that the Salk vaccine was "safe, effective, and potent." Most people were ecstatic, but, as is common in the face of great success, a few felt hurt for not sharing enough of the credit, and others were simply envious. Jonas Salk had "lost his anonymity" and became a popular hero overnight.[4] The Salk Institute for Biological Studies was in nobody's mind on that fateful day, although the now-famous institute never would have existed without the breakthrough announced at Ann Arbor. The Salk Institute owes its existence to Jonas Salk's success in developing the first effective polio vaccine and to his relationship with Basil O'Connor.

In 1955 O'Connor was the powerful founding president of the NFIP, later known as the National Foundation–March of Dimes (NF-MOD).[5] Preventing polio was a major public health problem in the United States and Europe in the 1950s. The disease's most famous victim, President Franklin Delano Roosevelt (FDR), initiated a solution and Jonas Salk worked it out. It was Basil O'Connor who engineered the solution by connecting the early challenge of FDR and the work of Salk. However, O'Connor's contribution to science and his collaboration with Jonas Salk did not end with the polio vaccine. He went on with Jonas to create the Salk Institute for Biological Studies in 1960. The institute was the second achievement that these two men dreamed of and realized together. In both cases, exceptional people and high drama were involved. This is the story of the second dream of Salk and O'Connor.

O'Connor had an extraordinary career as a venture philanthropist, and it may seem surprising that his biography has not been written, although he appears as a major character in many books and articles. The adjectives used to describe him may well account for the absence of a definitive biography: brusque, bombastic, blunt, abrasive, feisty, off-putting, with a low boiling point, and light in personal charm. Perhaps, however, those are some of the characteristics that helped him make history.

Daniel Basil O'Connor was born in Taunton, Massachusetts, on January 8, 1892, to a poor family of Irish descent. He won a fellowship to Dartmouth College, where, admitted as a freshman at the age of sixteen, he got the nickname that his close friends used throughout his life, "Doc." The name was in part a joke, because he was "the skinniest freshman ever seen," a slight boy weighing only 111 pounds, while the big Dartmouth football coach at the time was a John "Doc" O'Connor. However, the name also referred to his initials, since, although he was known as "Basil," his first name was actually Daniel—a name he dropped when he moved to New York City and discovered how many Daniel O'Connors were listed in the New York telephone directory.[6]

Supplementing his fellowship by playing the violin with a small dance band, O'Connor graduated from Dartmouth at nineteen. He was planning to teach for two years to save funds for law school, but another extracurricular activity again served him well: debating. Story has it that Thomas Streeter, a member of the Boston law firm Streeter and Holmes, was a judge at a debate O'Connor had competed in while at Dartmouth. He was so impressed by the student's performance that he offered him a loan that allowed him to enter Harvard Law School immediately.

After graduating from law school in 1915, O'Connor took a job for one year with the prestigious New York law firm Cravath and Henderson. By 1916, however, he had moved to Boston to work for his benefactors Streeter and Holmes, where he remained until 1919. As the United States entered World War I in April 1917, both Streeter and Holmes were in military service, while O'Connor was exempt because of his poor eyesight. This appears to have been an extraordinary opportunity for the young lawyer to study and practice law, and this period contributed greatly to his future success.

In 1919 he opened his own practice in New York at 120 Broadway. The main interest of his office was negotiating contracts between oil producers and refiners. There are several plausible versions of how O'Connor became acquainted with Franklin D. Roosevelt. Dartmouth president Ernest Martin Hopkins is said to have introduced the two young lawyers. They met again at the 1920 Democratic National Convention when O'Connor's brother John was serving in the New York State Assembly and Roosevelt was running for the vice-presidency. Although both O'Connor and Roosevelt occupied offices in the famous 120 Broadway building where they may have met, it was a client of O'Connor, an oilman named John B. Shearer, who knew Roosevelt personally, who brought them together in 1924.

The law partnership of Roosevelt and O'Connor was announced on January 1, 1925. O'Connor wanted the Roosevelt name on the firm to attract clients. Roosevelt, now seriously disabled and struggling with paralysis, wanted his name to remain visible as the main partner of a law office. This would also provide him with a small, steady income independent of his family fortune, whether he was able to work or not. The partnership ended when FDR became president in 1933,[7] but their close relationship lasted until FDR passed away on April 12, 1945.

In 1921, at the age of thirty-nine, FDR had been stricken by a paralysis of the lower body while on vacation at his summer home on Campobello Island, Maine. His condition was diagnosed as polio.[8] By 1924 he had fought back his disease for three years, enduring every treatment possible with little result while at the same time trying to minimize his condition in the eyes of the public.[9] In the summer of 1924 FDR heard from his friend George Foster Peabody about the benefits of warm spring treatments in restoring strength to partially paralyzed limbs. Peabody happened to be part owner of such a facility, Georgia

Warm Springs. FDR promptly visited the place in October 1924. That visit gave the desperate FDR renewed optimism, and, although he would never walk again, it was the first step toward a solution to the polio challenge.

Georgia Warm Springs, about eighty miles south of Atlanta, was a run-down resort with a shabby wooden hotel and a few flimsy buildings.[10] Its primary attraction was a large swimming pool fed with warm spring water high in minerals that made it buoyant. The resort and its pool made FDR feel so good and hopeful that he convinced himself that he had found the perfect place for rehabilitation, and he dreamed of restoring the location as a great health resort and a modern treatment center for polio. Against the advice of his wife, Eleanor, and of O'Connor, he bought the ramshackle resort in 1926. He turned it into a nonprofit organization, the Georgia Warm Springs Foundation, the precursor of the NFIP.

Although O'Connor was not initially interested in the project, it is while visiting FDR at Warm Springs that he became enticed to participate in the polio campaign. In 1928, when FDR reentered politics as the governor of New York, O'Connor found himself appointed treasurer of the Georgia Warm Springs Foundation. The laudable goal of the foundation was to provide rehabilitation to any victim of polio, but this aspiration was bound to soon bankrupt it, and the foundation almost failed during the Great Depression. When FDR was elected American president and took office in 1933, O'Connor, as treasurer of the foundation, was left holding the bag.

O'Connor involved a group of people interested in the Georgia Warm Springs Foundation to discuss possible fund-raising strategies. Together they came up with the idea of annual President's Birthday Balls. On January 30, the birthday of FDR, great balls organized by local committees were to be held all over the country. This was the first scheme to raise funds through a publicly supported and nationwide campaign. The strategy was much more successful than most people, including the skeptical O'Connor, had expected it to be. The first birthday balls in 1934 brought in more than one million dollars, which allowed continued support of the Georgia Warm Springs Foundation and the initiation of a nationwide program to fight infantile paralysis. However, linking philanthropic fund raising so closely to FDR was seen as highly political, and in 1938 FDR announced the creation of the nonpartisan National Foundation for Infantile Paralysis (NFIP), whose goal was not only to provide treatment to polio victims but also to find a cure for the disease. The committees for the birthday balls eventually evolved into the March of Dimes organizations and the funding arm of the NFIP. Later, the March of Dimes became part of the official name of the foundation itself that became known as the National Foundation–March of Dimes.[11]

The same year the NFIP was founded, in 1938, O'Connor appointed Thomas Rivers, a prominent virologist from the Rockefeller Institute, to lead the Medical Advisory Committee of the NFIP.[12] Rivers was regarded by his colleagues as a pioneer and leader in the relatively new field of virology. As such he was able to recruit some of the best researchers to participate in the work of his committee. In 1938 little was known about polio, and it was time to survey the field and clarify the gaps in knowledge. After disastrous early attempts to develop a vaccine it became clear that there could be no shortcuts: success would come from painstaking work and patience. Rivers and his committee of experts defined an agenda of basic research and got involved in establishing priorities and distributing funding. Another important early decision of Rivers's committee was to train young scientists in the field of virology by offering fellowships.

It should be pointed out that the National Institutes of Health (NIH) did not initiate a program of extramural research grants and fellowships until 1946, and the National Science Foundation (NSF) was not created until 1950.[13] It took some years after World War II before major support of medical research by government agencies became available. Until then private foundations such as the NFIP and the Rockefeller Foundation played a leading role in funding medical research. By supporting basic research and training and sponsoring virology conferences,[14] the NFIP helped build the fields of virology and molecular and cellular biology while developing a solution to the polio problem.

At the request of FDR, O'Connor had accepted the presidency of the NFIP in 1946. He soon recruited Harry M. Weaver as director of research. Although Weaver had himself done some work on polio, he was more talented as an organizer and a catalyst than a researcher. He overhauled the system for funding medical research[15] and should also be credited for attracting Jonas Salk to polio research and introducing him to O'Connor.

Jonas Salk was born in New York on October 28, 1914, in a modest but supportive Jewish family. The oldest of three sons, he took advantage of the educational system open to him and many other gifted Jewish boys. He graduated from the highly competitive College of the

City of New York (CCNY) in 1934 and earned his M.D. from the New York University (NYU) College of Medicine in 1939. Moreover, he supplemented his medical education with fellowships that introduced him to medical research. In 1938 Thomas Francis had become head of the Department of Bacteriology at NYU and he became Salk's mentor. Francis, a pioneer in the field of influenza,[16] had worked for several years at the Rockefeller Institute, where he had isolated the virus. In his lab at NYU he was working on an influenza vaccine. He believed that a killed virus could produce an effective vaccine. This was contrary to the belief of most virologists, who assumed that a natural infection with a live virus was required, even if it were only a mild infection by a weak or "attenuated" virus.[17] During his last year of medical school Salk worked with Francis on early experiments with killed influenza virus.

By the time he graduated Salk had distinguished himself not only as a gifted clinician, but also as a dedicated medical researcher. This won him a prestigious internship at Mount Sinai Hospital. When his internship ended in fall 1941, finding an attractive research position in New York proved to be a major problem. Thomas Francis had just moved to the University of Michigan as chairman of the Epidemiology Department, and Salk inquired whether a position might be available in his new laboratory. His only option was to obtain a fellowship to support an appointment with Francis, at least for one year. A strong letter of recommendation from Francis won Salk a National Research Council fellowship. However, in December of that year the United States entered World War II and Salk was to be drafted into the military. Again upon Francis's recommendation, Salk obtained an occupational deferment, and he arrived at the University of Michigan in early 1942.

Francis's lab was well supported by the army as well as a hefty grant from the NFIP. Salk spent six years in the lab with Francis, during which he assumed increasing responsibilities. By the end of the war, however, he was still only a poorly paid research associate, and he was getting impatient. In 1946 he finally became assistant professor, but by then he was exploring possibly better jobs elsewhere. He was confident that he had learned all he needed from his work with Francis: the principle of killed-virus vaccines, the use of formaldehyde to kill viruses and of adjuvants to stimulate the immune response, and the large-scale controlled human trials as practiced on soldiers for the influenza vaccine experiments. He was grateful to his mentor but eager to be on his own.

Eventually, in 1947, he heard of an opening at the University of Pittsburgh (Pitt), where the main attraction was independence. Pittsburgh was then a grimy and provincial city, but the community was wealthy and planning a renaissance. William McEllroy, the dean of the School of Medicine, was very supportive of research, but the academic position he had to offer was complicated. Salk would be hired as an associate research professor of bacteriology, an appointment shared between the School of Medicine and the Division of Research. The organization was "a bit loose," as the relationship between the Division of Research and the Medical School was "a little difficult."[18] It was unclear to whom Salk would report. This administrative confusion was perfect for Jonas because the absence of clear rules gave him a measure of control.

He ignored a few customary procedures by applying for an army grant for influenza research and by negotiating directly with Weaver for NFIP support. Eventually Weaver presented Salk and Dean McEllroy with a $200,000 annual contract, an offer that the university could hardly refuse. The contract was for classifying all available strains of poliovirus into immunologic types. Actually, there was already evidence that at least three types of poliovirus existed and that infection by one type did not confer immunity to another. This information was so essential that it required absolute confirmation. It would be unglamorous, time-consuming, routine work that more established polio scientists would balk at. It would be Jonas's liberation. Those funds would provide him with staff, space, and facilities that would be organized and headed by him. There was plenty of empty space in the old Municipal Hospital, at least in the basement. Jonas saw the opportunity to build his own lab from scratch, starting at the bottom. He would learn about polio in the process.

Since the 1940s it had been known that poliovirus entered the body through the mouth and appeared briefly in the blood after infection. There was a good chance, therefore, that an antibody might kill the virus before it spread from the intestines to the nerve tissues causing paralysis. After decades of confusion, the polio field was shaping up.[19] Then, in 1949, two breakthroughs occurred. First, Isabel Morgan at Johns Hopkins had been able to immunize monkeys by injecting poliovirus killed by formaldehyde. The same year John Enders and his colleagues in Boston had succeeded in growing poliovirus in test tubes on cultures of various types of nonneural tissues.[20] This meant that large quantities of poliovirus could be grown and easily purified to prepare potent vaccine, free of nerve tissue that could cause serious harm through an immune reaction. By 1950, while virus typing was underway, Salk wasted no time and set up test tube cultures of poliovirus in his lab.

In September 1951 the Second International Poliomyelitis Congress was held in Copenhagen. Salk reported on the virus-typing program: only three antigenic types had been found so far. This was solid, boring work, as expected from a newcomer to the polio field. However, the most important event of this conference for Salk was still to come. For the return trip from Europe Weaver had arranged passage for Salk on the *Queen Mary*. Weaver had mentioned Salk to O'Connor, but the two men had only met casually in Copenhagen. On the ship Jonas and Doc became friends. They took long walks on deck, enjoyed long talks, and discovered that they agreed on most important issues. Later Salk liked to summarize their relationship in practical terms, saying, "Basil O'Connor and I brought out the best in each other."[21]

Back in his lab after Copenhagen, Salk and his team worked like dogs to examine hundreds of variations to gear up for experiments in humans. In July 1952 Salk inoculated forty-three children—all polio survivors—with an experimental polio vaccine. Salk claimed this was not bravery: "It is courage based on confidence, not daring, and it's confidence based on experience." Yet he later admitted, "When you inoculate children with a polio vaccine, you don't sleep well for two or three months." The early human experiments were done in secret to avoid interference from publicity. Only O'Connor, Weaver, and Rivers were aware of Salk's experiments. By December 1952 the results were in: the antibody titers had risen and were holding at high levels. By then 161 people had been inoculated with various preparations of formaldehyde-inactivated poliovirus preparations with no obvious harm. It was time for more extensive trials and to present the results to the medical community.

Salk reported his results at a private scientific meeting in Hershey, Pennsylvania, in January 1953. The meeting participants were all NFIP grantees, including Albert Sabin and John Enders. Most polio researchers urged prudence and expressed skepticism. Sabin and even Rivers still doubted the effectiveness of the killed virus vaccine, and Enders recommended much more experimentation. Weaver informed the NFIP trustees of Salk's progress in a cautious speech on January 26. Then, somehow, the word got out. The newspapers stirred anticipation before the publication of the first scientific report, not due to appear for another two months. The people were thrilled but scientists were shocked; this was the first sign of Salk's future fame with the public and rejection by the scientific community. The reporting of scientific results by the media before their publication in a scientific journal was considered simply unacceptable.

To make matters worse, in February O'Connor called a special meeting in New York at which a number of dignitaries from the medical community and the press were invited. No polio researcher except for Salk was invited. Rivers presented the problem at hand: should large-scale human trials of the Salk vaccine be done at this stage? Salk presented his results. Performing large-scale trials received support in consideration of the fact that children would be dying while waiting for an ideal vaccine. The newspapers sensationalized the story, and Jonas knew then that he was in trouble. The public was made to believe that a vaccine was at hand and would be available for the next polio season, which was an impossibility. To try to contain the damage Salk made a serious mistake. He suggested to O'Connor that they appear together on the air to explain directly and honestly to the public the exact state of affairs of a polio vaccine. On March 26, two days before publication of the scientific report in the *Journal of the American Medical Association,* Salk and O'Connor talked on the CBS radio show *The Scientist Speaks for Himself.*

Clearly O'Connor, as president of the NFIP, was seeking publicity for his foundation. Some believed that Jonas was seeking fame for himself. His friends thought that he was simply naïve not to have realized the risk of such a public address to his scientific reputation. Actually, Jonas's radio speech was dignified and informative,[22] even though many questioned his motives. In truth, Salk did not want large-scale human trials to begin immediately. He was not quite ready, but he did not want to have to wait for many years either. To achieve the human trial stages, he needed the continuous generous support and the complete cooperation of the NFIP. He was badly caught between the standards of his fellow scientists, the public relations requirements of the NFIP in general, and the extreme pressures and expectations of O'Connor in particular. By accepting major NFIP support, Jonas had put himself in a difficult situation, but he had no choice at this point. Amazingly, he found himself in a very similar situation when he undertook creating the Salk Institute (see chapter 8).

As 1952 had seen the worst polio epidemic in U.S. history, there was a great sense of urgency. The results published by Salk in March 1953 were clearly encouraging: the vaccine boosted immunity with no serious side effects. While Salk was inoculating hundreds of volunteers in the Pittsburgh area, the NFIP was planning a large field trial of his vaccine. The foundation had chosen to test the Salk killed vaccine rather than wait for a live vaccine to come along. Effective preparations of the Salk vaccine were available, but obtaining the appropriate attenuated mutants from all three types of poliovirus necessary to prepare live vaccine could have taken years.[23] In summer 1953, the design of the field trials was being hotly discussed, and this led Harry Weaver to resign.[24] By December the plans for the mass national field trials were in place, and Tommy Francis once more endorsed Jonas by agreeing to oversee the field trials. Francis did, however, put some conditions on his participation. These included the use of injected placebo controls, a blank check but no meddling from the NFIP, and complete secrecy of the results until their analysis was complete. The Poliomyelitis Vaccine Evaluation Center was set up at the University of Michigan.

The year 1954 was to see what has been called the biggest public health experiment ever. It involved more than 300,000 doctors, nurses, and volunteers to inoculate 1.8 million schoolchildren at 217 sites in the United States. Two drug companies had prepared the vaccines that had been extensively tested. Begun in April 1954, the mass field trials ended in late spring, before the polio season started. It took Francis almost a year to evaluate the results. Meanwhile, O'Connor—with the help of nine million dollars—had already convinced six drug companies to prepare and stockpile the Salk vaccine. He was gambling that the trials would have favorable results and that the Salk vaccine could be licensed quickly, in time for the polio season. By March 1955 Francis was ready to write his final report, the contents of which were still secret. Selecting the date and place for the official announcement took some soul-searching. Finally all agreed on April 12, which happened to be the tenth anniversary of FDR's death, and on using the Rackham Hall auditorium of the University of Michigan in Ann Arbor. This arrangement acknowledged the essential roles of both the NFIP and the University of Michigan.

Unfortunately, the news release turned out to be a stage show, with a setting and atmosphere inappropriate for reporting the result of a very important—and costly—scientific experiment. The interested audience had become much larger than the scientific community: it included some 1.8 million children and their parents and many thousands of volunteers, not to mention millions who had contributed money. After years of research and months of highly publicized trials, the nation had waited too long to be dispassionate. Tommy Francis gave a thorough ninety-eight-minute report to some five hundred scientists and 150 reporters in front of sixteen cameras and a battery of microphones. The University of Michigan press release had been distributed to reporters only a few minutes before Francis's presentation. Its opening sentence read, "The vaccine works. It is safe, effective, and potent." Salk himself had only heard the news that morning over breakfast with Francis and O'Connor. Later in the day, when Salk went up to the podium, he received a standing ovation. As a star he gained millions of fans but as a scientist he lost the respect of many colleagues. The press release affair was, simply, in bad taste, and it was followed by sensational front-page newspaper reports. The NFIP was held responsible and Salk was blamed for going along with their publicity stunt. It certainly was not Jonas's idea. Was he flattered? Wouldn't you be? Did he have a choice? No. Once more, he was caught.

The NIH had asked that a committee of experts be available in Ann Arbor to decide on the spot about the licensing of the Salk vaccine by the Public Health Service.[25] That Licensing Committee never had a chance to review the report. Considering the excited mob—which included a gang of newsmen—waiting outside, the committee was under extraordinary pressure, if not duress. It took only about two hours to grant commercial licenses to five drug companies that had prepared the Salk vaccine for mass inoculation. What was ignored was that three of those companies had not provided the vaccine that had proved safe in the carefully controlled field trials. The committee had no way to quickly evaluate the safety standards of those companies. Inoculations started immediately. On April 25, two weeks after the Ann Arbor announcement, a child became paralyzed eight days after vaccination. In the following days more cases of polio were reported in vaccinated children and in people they had come in contact with. The source of the virus was quickly identified as Cutter Laboratories, and their vaccine was recalled within two days of that first report. The inoculations soon resumed without any further problem. Still, it was a tragedy that left 164 victims paralyzed and eleven dead.

The Cutter incident left an important legacy.[26] It led to much greater involvement of the government in regulating vaccines, first by a division of the NIH and later by the FDA. As this incident was the first public health emergency that required a national response, it also emphasized the importance of early detection of epidemics and of the role of the CDC. Perhaps the best-known consequence of the Cutter incident is the principle of "liability without fault." The parents of one of the victims sued Cutter Laboratories and, although the jury found that the company was not negligent, their verdict was that Cutter was liable anyway. This decision created an epidemic of lawsuits that for decades scared many drug companies away from manufacturing vaccines and created vaccine shortages. Only recently, with new technologies and protective legislation, has the vaccine industry begun to recover.[27]

Understandably, the Cutter incident caused fear and raised questions and doubts. This detracted from the popularity of the Salk vaccine and opened the door to attacks by the proponents of the live attenuated vaccine. In principle, a live vaccine might have advantages, including a faster and more lasting effect, and the convenience of oral administration was undeniable. However, attenuated viruses had been seen to reverse to virulence, making them even more risky than killed viruses. While several researchers were working on a live polio vaccine, the most vocal opponent of the Salk vaccine was Albert Sabin. By 1959 Sabin had made considerable progress with the generous support of the NFIP. The foundation had by then invested $1.3 million in his work but was not ready to support mass field trials of a live vaccine in the United States since the incidence of polio had declined steadily and remarkably since inoculations with the Salk vaccine had started.

Meanwhile, Sabin had established contacts with scientists in the Soviet Union, an extraordinary feat at the height of the Cold War. He provided the live virus strains to immunize ten million children in the USSR in 1959. Eventually almost eighty million young people were vaccinated with the Sabin vaccine in the USSR. This program, even larger than the 1.8 million inoculated in 1954 mass field trials of the Salk vaccine, cannot, however, be considered a public health experiment. Carried out without controls, it was more of a massive public health campaign. Perhaps miffed by being left out of the decisions concerning the Salk vaccine, the American Medical Association endorsed the Sabin vaccine for use in the United States in 1961, before it was even licensed in 1962. The oral live vaccine, cheaper and easier to dispense, was the vaccine of choice for developing countries, and the World Health Organization recommended it for the global eradication of the disease. By 1965 the Sabin vaccine had essentially replaced the Salk vaccine in the United States and in most of the world, except for Holland and Scandinavia, which persisted in using the Salk vaccine and remained polio-free.

By 1979 it had become clear that the Sabin vaccine was the cause of the remaining rare cases of polio in the United States. The wild poliovirus had been eliminated in this country, but, as expected, attenuated Sabin strains were able to revert to virulence. It took another twenty years, until 1999, for the United States to switch back to the Salk vaccine, which had meanwhile been improved by researchers in Holland. By then both Salk and Sabin had passed away. Unpleasant as the Salk-Sabin rivalry may have been, the NFIP was wise in generously supporting both programs. In the end, it is a combination of the two approaches that has the best chance to control polio worldwide today, provided that improved strains of attenuated virus are used and that they are engineered to be unable to revert.[28]

It is remarkable that Jonas Salk separated his polio involvements entirely from the Salk Institute, where few were aware of his vaccine problems. Actually the Salk-Sabin controversy and the Cutter incident combined, in a sinister way, to undermine the reputation of Jonas Salk and of the NFIP. This led to the major financial crisis of 1962 that almost ruined the plan to create the Salk Institute (see chapter 8). However, clearly the single most important event in the genesis of the Salk Institute was the fact that FDR was affected by polio. Or was he? His disease was diagnosed as polio in 1921, soon after he contracted it in Campobello. However, a 2003 scientific paper reexamined the cause of his paralysis.[29] FDR's age at the time of his illness, thirty-nine, and some of the features of his illness appear more suggestive of Guillain-Barré syndrome than of polio. A spinal tap could have distinguished between the two diseases, but none was done, and we will probably never know. A different diagnosis would not have changed the fate of FDR, since there was no cure for either disease. However, if it had been known at the time that FDR did not have polio, there would have been no NFIP, less pressure to solve the polio problem, and no Salk Institute. This raises the amazing possibility that the Salk Institute could be the result of a lucky misdiagnosis!

CHAPTER 2

Doctor Polio Meets Doctor Atomic

An institute sticking out all by itself has great advantage.[1]

—J. Robert Oppenheimer

In the summer of 1955, in the midst of the excitement over the polio vaccine, the University of Pittsburgh (Pitt) was in a state of flux—and expectation—as negotiations to appoint a new chancellor were underway. Chancellor Rufus Fitzgerald had reached age sixty-five and, at his farewell dinner, he had advised the new administration "to bring the University of Pittsburgh into the forefront of American institutions of higher education." The trustees agreed and had found their man: his name was Edward Harold Litchfield, and he became the twelfth chancellor of Pitt.[2]

Litchfield was then dean of the Graduate School of Business and Public Administration at Cornell University. He looked at administration as a profession in itself. He thought that a competent manager of any organization, whether a church, regiment, hospital, or prison, should have no problem administering a university. Not surprisingly, many disagreed with that view.

Born in Detroit, and with a B.A. in political science and a Ph.D. in public administration from the University of Michigan, Litchfield at age forty-one had already been successful in several positions. However, the position that seems to have influenced him most was that of advisor to General Lucius D. Clay, who succeeded Eisenhower as the U.S. military governor of Germany in 1947. During those difficult postwar times there were a variety of challenges in occupied Germany.[3] Clay's style was that of a man insisting on order and a tight organization, and it was said, "He looks like a Roman emperor—and acts like one." He left his mark on Litchfield.

Litchfield had been discovered in the summer of 1954 by J. Steele Gow, an officer of the Falk Foundation, to which Litchfield had applied for a grant. Gow was the right-hand man of Leon Falk, who was a trustee of Pitt and the chairman of the chancellor search committee. After some exchange of correspondence, Litchfield visited Pitt and met the committee members, including the chairman of the board of trustees, Alan Magee Scaife. The trustees saw in Litchfield everything they wanted, including not only experience in administration and government but also contacts with business that are essential in an industrial city such as Pittsburgh. Litchfield saw in Pitt a great opportunity for an expert in administration because it was said to be the "most under-administered institution in American education."

Eventually Litchfield sent to Scaife and Falk an extensive statement of his understanding of their goals. It was nothing less than an ambitious program to turn Pitt into one of the foremost universities in the world at the cost of "many millions of dollars over the next decade." His election as the twelfth chancellor was announced at a press conference on July 18, 1955, but he was not to take office until almost one year later. From July 1, 1956, Litchfield divided his time between his many different positions. Then in August he went on an extended trip abroad, and, immediately after his return, he accepted the position of chairman of the board at Smith-Corona. At that point his supporters—Scaife, Falk, and Gow—expressed uneasiness about the time and effort the new chancellor was dedicating to activities outside the Pitt campus. To counter their concerns and to cope with his travels, in the fall of 1956 Litchfield bought his first private plane, a twin-engine Apache that he replaced one year later with a Beechcraft of greater capacity.

Other characteristics of the new chancellor were a source of gossip and mild shock among the Pitt community. He had a strong sense of the ceremony owed to the chancellorship. He wished to be escorted to all meetings on campus and expected everyone to rise when he entered a room as a sign of respect for his office. For the three-day inaugural celebration, on May 10, 1957, he had a gown specially designed of antique gold faille of the best quality with blue velvet stripes on the sleeves (the colors of Pitt), and he wore, instead of the conventional square flat-top hat with a tassel, a striking velvet beefeater beret. Moreover, he had a chancellor medallion crafted, held on a heavy gold chain that Scaife put around his shoulders at the ceremony. Within a year of the inauguration he ordered an elaborate covered platform to be erected for all major outdoor events. It resembled the fancy medieval canopy under which the king watched jousting tournaments, and it probably earned him his sobriquet "King Edward." Not quite a Roman emperor but close enough! His flamboyant style bordered on the ridiculous, but, combined with his boundless energy and ambition, it attracted curiosity and the attention of journalists and resulted in colorful articles in numerous magazines.[4]

Some Pittsburghers were offended by the criticisms voiced by Litchfield and his team about the state of affairs at the time of their arrival. However, everybody praised the new chancellor's goal to raise academic standards and boost research. The success of the Salk polio vaccine put Pitt in the limelight. Several members of the Pitt public relations staff were needed to handle the publicity. Mail poured in, much of it containing money. Since Salk's arrival at Pitt in 1947, his laboratory had been housed in rented space at the Municipal Hospital of the City of Pittsburgh. Centrally located in the Pitt Health Center, the Municipal Hospital was originally designed for the quarantine of contagious diseases and was the central facility for the care of polio victims in western Pennsylvania. By 1957, with the decline of contagious diseases in general and of polio in particular, much of its space stood empty. In May 1957, almost immediately after Litchfield's inauguration, Salk was informed that funds had become available to the university for the purchase of the Municipal Hospital in the form of a gift from donors who wanted to remain anonymous at the time. Their wish was for the building to be renamed Salk Hall, honoring Jonas Salk and making available to him space to pursue his work at Pitt. This provided an opportunity for growth and for the creation of the first Pitt Research Institute around the name and fame of Pitt's hero, Jonas Salk.

The first mention by Jonas Salk of an institute being planned in Pittsburgh appears in notes recorded on May 25, 1957, and transcribed by his devoted secretary, Lorraine Friedman.[5] Lorraine had worked for Salk since 1948. This was a time when administrative assistants were proud to be called secretaries because it suggested that they shared some of the secrets of their bosses. Salk clearly trusted Lorraine, confided in her, and valued her judgment. He recorded very spontaneous drafts of letters—some never to be sent—and confidential thoughts and plans, and it is touching to see at the bottom of some of these transcripts a large, scribbled handwritten note he added upon proofreading,

"L. WHAT DO YOU THINK?" These remarkably candid and revealing notes are precious sources of historical information.

His 1957 notes already included important concepts for the institute as envisioned by Salk at the time. He proposed the Institute for Experimental Medicine, where original observations would be extended to develop solutions to problems arising from diseases or from man's relationship to man. The first areas of interests would be virology and immunology. He considered it important for such a research institute to have close professional ties with a university. Then he mentioned the foremost issue, that "the administrative and financial structure should be independent." His notes also introduced the original group of Pitt faculty members that Salk had in mind to start the institute with him: Thaddeus (Ted) S. Danowski and Frank J. Dixon Jr.

Danowski had been a professor of medicine at Pitt Medical School since 1948. He was a distinguished endocrinologist actively involved in research and a pioneer in developing patient self-monitoring of blood glucose for the management of diabetes. Dixon was chairman of the Department of Pathology. Salk, as leader of the search committee to appoint a new Pathology Department chairman in 1951, had been instrumental in Dixon's selection. Both Danowski and Dixon, like Salk, had vigorous research programs, and they raised considerable extramural financial support. Like Salk, both were members of the Division of Research and had been hired by Dean McEllroy, the codeveloper of the Folin-McEllroy test for urinary sugar. A medical researcher turned administrator, he wanted to encourage more research in the medical school.

Salk's notes simply stated that, by mutual consent, Drs. Danowski, Dixon, and Salk were desirous of joining together to establish the beginnings of an institute that would provide a university setting in which they could engage in full-time research: Dixon in the field of immunopathology, Danowski in the field of metabolic and endocrinologic disorders, and Salk in the field of viral diseases and neoplasia. Younger individuals and visiting professors would be brought in, and undergraduate and graduate training would be offered in the fields of experimental medicine represented at the institute. It would be an institute for basic research, but, wherever possible, it would "translate to the clinic" findings that might be of value for the prevention or cure of disease.[6]

Discussions and negotiations went on for more than two months, yielding a proposal developed through many drafts and that eventually gained the approval of Gow, A. C. Van Dusen, and the chancellor. The proposal was forwarded to Salk on August 21. This document brings up the second problem the organizers faced: the name of the institute. When first mentioned in the text the full name used was The Salk Institute for Experimental Medicine in the School of Medicine, abbreviated throughout the document as The Salk Institute. That the name was a real source of contention is evident from successive drafts of various memos exchanged between Salk and Litchfield dated September and October 1957, which show the name Salk crossed out then added back. Finally, in a memo for the record dated October 16, Litchfield spelled out, "We both agree that it would be a little awkward for Dr. Salk to raise funds for an institute that bore his own name. His preference and mine would be that for the time being the institute have a more general title and that it become known as the Salk Institute only when he relinquishes the directorship." A revision of that memo two days later stated that the institute would be known as the Jonas Salk Institute for Experimental Medicine. Dr. Salk would remain on the Pitt faculty as the first Commonwealth Professor in Preventive Medicine, while an administrator would be director of the institute. By November, Litchfield raised problem number three: the need for funding to remodel the building and for operating costs above the direct costs for research. The chancellor suggested that perhaps a foundation should be established to create and maintain the institute.

Litchfield knew that the situation was at a standoff and did what was expected: he set up an advisory committee. He accepted the suggestions for membership made by Salk, and on December 19 he sent out letters of invitations. All of those invited accepted, and in a press release dated January 18, 1958, Litchfield announced the appointment and composition of the advisory committee. The outsiders included J. Robert Oppenheimer and Tom Rivers as well as Dr. Lowell Reed, an advisor to the chancellor for health affairs, and two representatives of industry, Mark Cresap, president of Westinghouse, and Fred Foy, president of the Koppers Company in Pittsburgh. Members of the Pitt faculty included, in addition to Salk and Dixon, Dean McEllroy, who would serve as committee chairman, and Hubert Bloch, chairman of the Department of Microbiology. Chancellor Edward Litchfield (ex officio), Van Dusen (assistant chancellor for planning and development) and Lucien A. Gregg (assistant to the chancellor for health affairs) would represent the Pitt administration.

Most prominent on that committee was Oppenheimer, "the father of the atomic bomb," who, after working as director of the Los Alamos Laboratory during World War II, had returned to Berkeley until 1946, when he became director of the Institute for Advanced Study at Princeton.[7] By the time of the January 1958 press release, an organizational meeting had already been held. Litchfield asked to see where the new institute would fit into the organization of the university. Oppenheimer had retorted, "Dr. Litchfield, we don't need any table of organization. That's why the United States doesn't have any Sputniks in the sky!" The first Sputnik was launched in the Soviet Union on October 4, 1957. This development had caused American dismay, anger, and frustration. Many, including Oppenheimer, were blaming bureaucracy for the failure of the United States to win the space race. Indeed, while the air force, navy, and army were still arguing over who should build the rocket to launch a sophisticated satellite, the Soviets had simply rushed ahead and launched a small, simple satellite, thereby establishing a proof of principle and beating the Americans into space.

Oppenheimer had confronted attempts at bureaucratic organization at Los Alamos, where an experimentalist, John Manley, had acted as his right-hand man in setting up the laboratory in 1942. Manley had urged Oppenheimer to develop a clear organizational structure but found him entirely unresponsive until he visited Oppenheimer again in January 1943. At that time, Manley writes, "I had scarcely opened the door when he shoved a paper at me, saying 'Here's your damned organization chart.'"[8] The organization at Los Alamos turned out to be based on personality rather than a bureaucratic structure: Oppenheimer established his authority by his omniscience, his omnipresence, and his pure charisma.[9]

Oppenheimer not only resented bureaucracy but also hated formality and the decorum so dear to Litchfield. While director of the Los Alamos Laboratory he used to identify himself on the phone by simply saying, "This is Oppie," and, in a meeting where an army officer had complained of the disrespect of a young enlisted engineer who had sat on the edge of his desk, Oppenheimer replied, "In this laboratory, anybody can sit on anybody else's desk." That rule applied to all, including the director, and a young scientist found that "his office was always open and each of us could walk in, sit on his desk, and tell him how we thought that something could be improved."[10]

Oppenheimer had also expressed doubts about the usefulness of advisory committees in general, and it is surprising he agreed to serve on this Pitt committee, especially considering the serious problems he had faced earlier. In 1954 the Atomic Energy Commission had withdrawn Oppenheimer's clearance after painful and humiliating national security hearings.[11] Various people have been blamed for Oppenheimer's disgrace, from Joseph McCarthy, the infamous senator from Wisconsin, to Edward Teller, who wanted to be the father of the H-bomb, to Teller supporters Lewis S. Strauss and Ernest Lawrence. The hearings left Oppenheimer a broken man who had lost his consulting posts in Washington and some of his self-confidence. Strauss, a trustee of the Institute for Advanced Study, sought to have Oppenheimer fired from the institute as well, but his dismissal was avoided when the entire faculty sent a statement of protest to the board of trustees in June 1954.[12]

Before the Pitt committee meetings Oppenheimer and Salk did not know each other well. There is no evidence of any correspondence between the two before January 14, 1958, the date of Salk's first "Dear Robert" letter.[13] That letter, which concerns arrangements for a visit of Oppenheimer to Pitt in January 1958, ends, "It will be nice seeing you again," indicating that they had met before, probably on an earlier visit of Salk to Princeton. Although Oppenheimer must have had little affinity for Litchfield, one can see similarities between Salk and Oppenheimer: both were scientists who had become popular heroes, both had made powerful enemies among colleagues who made it a point to humiliate them, and neither would receive the ultimate recognition, the Nobel Prize. After the Cutter incident (see chapter 1), in which a defective batch of polio vaccine had caused almost two hundred cases of paralysis and killed ten people, both would forever know the guilt of being responsible for deaths. Of course, both men also saved countless lives, whether by preventing polio or by making unnecessary the invasion of Japan by the Allies. It is probable that Oppenheimer felt a kinship for Salk, aware that Salk would suffer from his achievements and be tormented by jealous people who had not achieved as much.

At the first meeting of the Pitt committee on January 17, 1958, Oppenheimer was absent, but Litchfield attended ex officio. The minutes of the first meeting report that the chancellor charged the committee to advise him on the organization of the Institute for Experimental Medicine, which would be established within the framework of the University of Pittsburgh.[14] He also reminded the committee that this was an attempt to establish a pattern for the formation of other research institutes, and that this first institute should be a model to follow.

After that meeting it was recommended that Salk meet individually with Oppenheimer and with Rivers. A brief note phoned in on January 22 followed by a confirming telegram indicates that Salk visited Oppenheimer at Princeton on January 29. Clearly this meeting was to discuss what transpired at the first meeting and to prepare for the second, which was set for February 22. Salk drafted several documents for consideration at the second meeting,[15] including an organizational chart as requested by Litchfield. Always thorough, Salk did his homework and studied the organization of seven entities already established in association with Pitt and drew a chart for each of them. He ended up with not one but, as Oppie would say, "eight damned organization charts" after adding one of his own design. In addition to these organizational charts, Salk submitted statements that included no concrete suggestions other than to establish a foundation "with some such designation as The Salk Foundation for Advanced Studies in Medicine" to support the Institute for Experimental Medicine.

The second meeting was recorded, and the transcript gives us a rare and fascinating glimpse of the dialogue that took place.[16] It offers insights into the personalities and styles of the protagonists, records the language they used, and reveals much about their relationships, their opinions of each other, and the amazing fame of Jonas Salk at the time. Clearly the two dominant figures on the committee were Oppenheimer and Rivers. Although both supported Salk, their styles were very different. Oppenheimer offered concrete and conciliatory suggestions, while Rivers, in contrast, had a confrontational and cranky manner.

It is Oppenheimer who pointed out that the circumstances of the founding of a Salk Institute were so special that there was little chance that it would start a pattern. The name of Jonas Salk was not only known in the four corners of the world, but it would attract money relatively easily. To illustrate the advantages of an institute that is not part of a university, Oppenheimer compared the Institute for Advanced Study at Princeton with the Radiation Laboratory at the University of California, Berkeley, where he worked for twenty years. At Berkeley, the Radiation Lab was attached to the Physics Department, and its members were privileged compared to the faculty in physics. This resulted in tensions and all kinds of complications. The Institute for Advanced Study at Princeton, in contrast, had been set up independent of the university. Its members had a different lifestyle, with more freedom and a higher standard of living. If it were part of the university, it would have been an agonizing problem to decide who should be promoted into this more attractive situation.

Oppenheimer considered a small and independent institute sticking out all by itself to have great advantages. The proximity of a university would, however, be essential. It would provide a rich academic environment and promote collaboration and the sharing of facilities. People would accept university appointments because of the presence of the institute and vice versa. On the other hand, Oppenheimer clearly considered that the administrative machinery of the university could hamper the running of the institute. He identified Salk's objections to Litchfield's concept: having to take decisions to people who would not be familiar with the situation at the institute, being frustrated by delays, and wasting time arguing and waiting. In other words, he expressed concern about the red tape that would result from a university affiliation.

Rivers had other worries. He wanted to know when the idea of an institute came about. Was it before or after the money for the building was offered? Salk remembered talking about the idea with Litchfield and Van Dusen in "November or December a year ago,"[17] though the money for the acquisition of the building became available only the following May. Rivers insisted that Salk should not play an important part in raising money for the institute, and that a Salk Foundation should be formed that, according to Rivers, could take Salk anywhere he might want to go: Ohio, California, or even Princeton. Rivers argued that any university would receive Salk with open arms because, he said, "Salk, you are the best-known doctor in the world, and people idolize you." Like Oppenheimer, Rivers insisted that fund raising for a research institute at Pittsburgh University would bring little money, while raising funds for a Salk Institute or a Salk Foundation would be quite easy.

It is Oppenheimer who brought up major characteristics of the planned institute as he saw it: no matter what its relationship with the university turned out to be, this should be an endowed institute. The committee had estimated that the endowment should be at least ten million dollars, be available before the institute started, and be supplemented by soft grant money. This would be an institute for a group of men, a group that would grow slowly at a rate determined by the science. On the other hand, there would be a group of citizens—the foundation—that would raise and administer the funds to support the institute. Jonas Salk should have no connection with that foundation, though his name would be used. As usual for a foundation, it should have a board of trustees. However, the institute should also have incorporation papers and have a board of trustees of its own.

With respect to the institute faculty, Oppenheimer pointed out that the initial choices would be critical, and that this was an issue that Dr. Salk had already deeply considered. At the beginning there would be three to five lifetime appointments. These appointees would be equivalent to the permanent members at the Princeton Institute for Advanced Study, and their positions would be secured by the institute endowment. Also helpful would be an advisory committee, a rotating committee of outsiders, the very best people who would meet with the institute faculty. Moreover, the faculty would essentially run the place, deciding what had to be done and recommending new faculty members. However, it was understood that matters involving finances would have to be taken to the board of trustees.

The full report of the committee was submitted to the chancellor around mid-March.[18] However, there is no evidence that an agreement was in sight at this time, and the discussions and negotiations continued. The results of the advisory committee's deliberations were as expected: nothing much was achieved except that Salk received a crash course in setting up an institute. Salk listened carefully, and later he would closely follow the plans laid out by Oppenheimer at that meeting, in many cases down to the specific numbers. The Salk Institute would eventually be founded by—in addition to Salk—four distinguished faculty members with lifetime appointments, it would have an endowment of ten million dollars and the support of a foundation, and the faculty would essentially run the place. Only it would not be in Pittsburgh, the foundation would not be a Jonas Salk Foundation, and it would take several more years to become a reality. The board of advisors would, indeed, be the very best people.

Meanwhile, with the success of the polio vaccine, the National Foundation for Infantile Paralysis had outlived its purpose and was in the process of reinventing itself. In 1958 it broadened its goals and changed its name to the National Foundation–March of Dimes, and it has remained a successful philanthropic organization known today as the March of Dimes Birth Defects Foundation.[19] During the period of the development of the polio vaccine, Salk had established a close relationship with the National Foundation's founding president, Basil O'Connor, who became equally involved with the institute project (see chapter 1). In September, six months after the committee report had been submitted, Salk sent a letter to O'Connor containing material that Oppenheimer had given him concerning the Institute for Advanced Study.[20] "I thought it would be well for you to read this before you might see him. I think your suggestion about a visit with him is an excellent one." Salk probably reasoned that a visit to the charismatic Oppenheimer and the institute in beautiful Princeton were likely to seduce O'Connor into supporting the idea of his institute to the point that O'Connor might broaden the goals of the National Foundation to include support of the Salk Institute in the making.

Abraham Flexner had established the Institute for Advanced Study in 1930, and he directed it until 1939.[21] Born in Louisville, Kentucky, Flexner was essentially an autodidact until, in 1884, at the age of seventeen, his oldest brother, Jacob, sent him to Johns Hopkins University, which had recently opened its doors and was the only modern graduate school in America at the time. The university's first president, Daniel Coit Gilman, had recruited outstanding teachers who were inspired educators and excellent researchers. Flexner admired Gilman and said of him, "He knew that there are men who teach best by not teaching at all." Flexner eventually used Hopkins as a model for the Institute for Advanced Study. After obtaining his bachelor's degree, he returned to Kentucky, where he opened his own college prep school. His students lived by the rule "No duties, only opportunities," a motto that would later prevail at the Institute for Advanced Study. He closed his "Mr. Flexner's School" after fifteen years to study at Harvard for a degree in psychology, but he found the work disappointing and dropped out. He undertook a critical investigation of higher education in the United States based on his own experience and wrote a severe evaluation of the system in *The American College,* his first book. As a result, the Carnegie Foundation for the Advancement of Teaching hired him to investigate medical education. He took as models two medical institutions that he trusted and to which he had access: the School of Medicine of his dear alma mater, Johns Hopkins, and the Rockefeller Institute for Medical Research, where his brother, Simon Flexner, was director. He inspected medical colleges throughout the United States and Canada and published his famous *Flexner Report,* which led to major reforms in medical education.

As an expert in education, Flexner became executive secretary of the General Education Board (GEB) funded by John D. Rockefeller. In 1929, after his retirement from the GEB, he was contacted by legal representatives of Louis Bamberger and his sister, Caroline Fuld, who had become multimillionaires by building a chain of department stores that were purchased by Macy's in 1929, shortly before the stock market crash. They wanted the advice of Flexner in endowing a medical college in New Jersey, the state where they had made their fortune. However, Flexner had a vision for a true community of scholars, a kind of hybrid university in which teachers and students would work together to advance knowledge. He suggested an organization called the Institute of Higher Learning or Advanced Studies and chose Princeton as its location. In his autobiography he describes the purpose and characteristics of the Institute for Advanced Study as follows:

> An institute such as I contemplated ought to be small and plastic; it should be a haven where scholars and scientists could regard the world and its phenomena as their laboratory, without being carried off in the maelstrom of the immediate; it should be simple, comfortable, quiet without being monastic or remote; it should be afraid of no issue; yet it should be under no pressure from any side which might tend to force its scholars to be prejudiced either for or against any particular solution of the problems under study; and it should provide the facilities, the tranquility, and the time requisite to fundamental inquiry into the unknown. Its scholars should enjoy complete intellectual liberty and be absolutely free from administrative responsibilities or concerns.[22]

The Institute for Advanced Study was incorporated in May 1930. The Princeton location was chosen not only because it is in New Jersey, but also because it is at a reasonable distance from important cities without being subject to big-city pressures. Also, Princeton University already had an excellent mathematics department and a good library. The institute began to function in 1933, temporarily located in a building made available by the university. Flexner knew that its success or failure would depend on the founding faculty he would be able to recruit, and he was lucky: in 1933 Albert Einstein, escaping from the growing troubles in Germany, would become the institute's first permanent member.

It is understandable that Salk and O'Connor were attracted by both the philosophy of the Princeton institute and some of its other characteristics, especially the relatively large number of temporary members and its broad interest in many fields. However, a major difference between Princeton and the institute envisioned by Salk is that the Institute for Advanced Study was not set up as a center for experimental research, and, with one exception, it had no laboratories. That made it much less expensive to build and to run. Moreover, the Princeton institute started with a considerable endowment that has been preserved, while the modest endowment that established the Salk Institute quickly vanished (see chapter 8).

In January 1959 Salk had not yet entirely given up on Pitt, and he sent a letter to Oppenheimer mentioning that he had been "in conversation with several persons" willing to help establish a research institute independent of Pitt.[23] For the first time he mentions that, although he would prefer staying in Pittsburgh, he would be able to move with "the support I now have."

Several notes and a telegram indicate that Salk did visit Princeton at the beginning of February, and, following that visit, he wrote to Oppenheimer, "The idea of not looking forward to a future here has lifted a great weight that has been bearing heavily upon me."[24] It has been reported that Oppenheimer asked Salk, "Did it ever occur to you to go to California?"[25] This could have happened at that last meeting in February 1959.

It soon became clear that the institute planned by Salk would be bound by University of Pittsburgh policies, salary scales, and hiring procedures, that its finances would be controlled by the university, and that the entire operation would have to put up with the bureaucracy designed by Litchfield. Dixon, like Salk, objected to the red tape and control by the chancellor, and both men eventually left Pitt for La Jolla. In 1961 Dixon moved his Pathology Department to the Scripps Clinic and Research Foundation, while Salk left in December 1962 to start his own research institute in the same area. At the time La Jolla was an essentially unheard-of suburb of San Diego, but much was happening there that would eventually put it on the map (see chapter 5).

As for Litchfield, he made ambitious reforms that resulted in an unprecedented financial crisis for the university in 1965. He was forced to resign from Pitt and later turned to his business affairs. His career was cut short when, on March 9, 1968, his private plane crashed in Lake Michigan, killing him and his entire family.[26]

CHAPTER 3

Enter Leo Szilard

He [Szilard] tends to overestimate the role of rational thought
in human life.

—Albert Einstein[1]

By September 1958, six months after the second University of Pittsburgh (Pitt) committee meeting, the negotiations with Litchfield were
bogged down and Salk was clearly thinking about alternatives. He had
kept in close touch with Basil O'Connor, the powerful founding president of the Polio Foundation and his strongest supporter, who continued to encourage Salk. Although one can only imagine Salk's conversations with O'Connor, it appears that O'Connor gave Salk indications
that the Polio Foundation might be willing to support the creation of an
institute independent of the university.[2] Moreover, conversations with
Oppenheimer in early 1959 further encouraged Salk to consider leaving
Pittsburgh and locating his institute elsewhere.[3] This brought up the
question of selecting members of the planned institute, if not colleagues
from Pitt. Already in September 1958 Salk had sent O'Connor a letter
reminding him of a proposal made by Leo Szilard in January 1957.

26 September 1958

Dear Doc:
    You may recall our discussions of Szilard's proposal for two
inter-dependent research institutes, operating in the health area. This
was almost two years ago. I think it would be well, in the light of
our recent discussions, to review the two memoranda that are
enclosed. . . . This may all take on a different light as a result of the
passage of time and the occurrence of other events. I would like to
discuss the substance of these with you . . . and, also, speak to you
about the question of your speaking to Szilard yourself. He is a strange

person but most unusual; he has a very keen mind although he is notorious for being impractical. He has a very fertile imagination and acts like a bee whose function in life is to spread pollen. He himself was eager for such an institute to be developed so that he would have a place, either to visit or to have as a home base where he would spend his time stimulating and cross-pollinating young and eager minds. You may recall, he was the one who was supposed to have written the letter Einstein signed and sent to Roosevelt suggesting the possibility of the atom bomb. You probably know about this first-hand as to whether or not it is true. He and Fermi were the effective team that worked together in Chicago.[4]

Who was Szilard, and what was his proposal about? Leo Szilard was an extraordinary Hungarian.[5] He was born Leo Spitz on February 11, 1898, in Budapest—then in the Austro-Hungarian Empire—and was the oldest of three children: Leo, his brother Bela, and Rose, their sister. When Leo was two years old his father, Louis Spitz, had Magyarized his family name to Szilard. The paternal ancestors of the Spitz family were eastern Jews who had settled in Slovakia. Leo's great-grandfather was part of the family legend, a somewhat nomadic shepherd who had only one son, Samuel, Leo's grandfather. Samuel took up farming on a leased estate that included a decaying old castle where he raised his ten daughters and four sons. Whether because of bad luck or poor management, Samuel's endeavors usually failed. He eventually moved his family to Kremnica, a small medieval town in central Slovakia, and he ended up bankrupt and bitter.

In spite of the Spitz family's financial troubles, all four of the sons attended Jewish elementary schools, where instruction was in German, and at age ten Louis was sent to high school in Kremnica. That is where he had to learn Hungarian, a strange language that he never spoke well. A poor student, he graduated from high school at the age of twenty, at which point he left Kremnica for Budapest, or, rather, Pest. Only seven years earlier the city of Buda, on the hilly west side of the Danube, had been united to flat Pest, to the east. There he studied engineering at the Institute of Technology, where, in spite of his language difficulties, he became a top student through hard work and sheer perseverance. In addition to his studies, Louis had to support himself by tutoring high school students in the German language and literature. He graduated as a civil engineer and was offered attractive jobs by large building contractors. Soon he started his own company and became wealthy planning and building bridges and new railway lines.

Louis Spitz and Tekla Vidor met in Budapest during the 1896 Christmas season, and they were engaged only one month later. Tekla was twenty-five, ten years younger than Louis. A serious girl, she admired Louis, a mature self-made man with broad interests that made for interesting conversations. In truth, she was also attracted by his gorgeous suntan, the result of his working outdoors. All of this made him so unlike the pale and boring young men she usually met. After the civil marriage ceremony in April 1897, a royal notary presented Louis with a marriage contract that revealed that Tekla had a considerable fortune of her own. Next came the Jewish ceremony at a synagogue arranged by Tekla's father, Dr. Sigmund Vidor, a well-known ophthalmologist and president of the Medical Association. The couple moved into an apartment in the Garden District, where Louis set up his contracting office, sharing it with Tekla's brother Emil, a young architect. Two different cultures blended in that neighborhood of Budapest: wealthy merchants, mostly Jewish, and Magyar aristocrats. This was the time when the daughters of rich Jews married Hungarian counts while their parents made generous loans to the noble families. At the same time, the Magyar court was granting titles of nobility to the richest merchants and including them as members of the parliament's House of Lords or as civil servants. As this was before anti-Semitism appeared in that area, Louis and Tekla's assimilation into that society was quite straightforward.

This is where Leo was born, and Tekla, having experience taking care of younger siblings and her sister's babies, had no problem tending to Leo. Tekla and Louis, however, had very different ideas about raising children, which generated friction and arguments. Louis was tender and spoiled Leo, while she was stricter and more puritanical. Leo's father spent as much time as possible with the toddler, often taking him for strolls in the park on Margaret Island in the middle of the Danube. On those occasions he noticed that Leo was remarkably inquisitive, asking not only his father but also the passengers on the ferryboat questions about everything. However, Leo quickly learned to become argumentative, contrary, and increasingly strong-willed. He grew to be difficult, selfish, arrogant, and rude—even to his grandparents Vidor.

By 1902 a major change took place in the Szilard family that would allow Leo to develop his personality. By then Bela and Rose had been born, and Tekla's sister had young children as well. Tekla's parents commissioned their son Emil, the budding architect, to build a beautiful house large enough for the entire Vidor clan to live under one roof. The

Vidor Villa was meant to serve as a model home to display Emil's talents and attract new commissions. It turned out to be a fabulous house, a five-story fairy-tale mansion in art nouveau style, with stained-glass windows, porches, turrets, and a curved staircase with an ornate wrought-iron handrail.[6] The Szilard family moved into the villa in 1902, when Leo was four years old, and he found himself surrounded by half a dozen playmates, including four cousins.

This was the situation in which Leo grew up and developed his character. He was soon making up all the rules of the children's games and then twisting those rules to justify breaking them while showing off how clever he was. Although he directed the games, he did not participate in them himself; he only took credit for the ideas. He had neither the manual dexterity and physical skills nor the self-confidence of his playmates. He never learned to swim or ride a bicycle, nor did he learn to drive a car later in life. He was squeamish and fearful, and he preferred sitting and talking with the governess or reading a book to exercise or sports. The children learned German, French, and English from their governess, while they spoke Hungarian with their nanny. A private tutor came to the Vidor Villa daily and taught them other subjects. Tekla closely followed her children's education while Louis encouraged them to eat lavish meals supplemented by unlimited delicacies and sweets. This is where Leo developed the inability to control his cravings and the unhealthy habits that would shorten his life. From an early age, his favorite pastime was to just sit, think, and enjoy wonderful ideas, whether his own or others'.

By 1908 his home tutoring ended and Leo enrolled in an eight-year high school. Under the influence of his father he chose a "practical" school to learn science and technology, planning to become an engineer.[7] It is in high school that he became intrigued by physics. He performed superbly in science classes but was handicapped by his sloppy drawing and physical clumsiness. Religion was not his forte either—understandably so, since both families, the Szilards and the Vidors, were nonbelievers, and neither practiced Judaism. He graduated from high school with highest honors in June 1916, and in September he started courses at the Palatine Joseph Technical University, enrolling in civil engineering. With girls he was distant but polite, using humor to mask his discomfort and escape emotional torments.

In the summer of 1914, however, when Leo was sixteen and the Szilard family was in Austria on vacation, Austria-Hungary allied to Germany had declared war on Serbia, starting the First World War. The

Szilards rushed back home to Budapest. Leo recalled their trip to Vienna because he noticed trains loaded with soldiers, many of them drunk. As someone commented on the enthusiasm of the troops, Leo retorted that this was not enthusiasm but drunkenness. His parents quickly pointed out that this was a tactless remark, and Leo later claimed that it was this incident that made him choose truth over tact for the rest of his life. It appears that Leo never understood that often it is not the truth that hurts but the way it is told. It was more than just verbal clumsiness in his case; he was entirely addicted to reason and pushed pure logical thinking to the bitter end, sometimes with tragic consequences.[8] He never considered other people's feelings and had only contempt for emotions, as he believed that they hamper rational thinking. He avoided close relationships to prevent obscuring his own reason, although he was not insensitive or malicious.

Back in Budapest, Louis discovered that the value of his assets was quickly dropping. This marked the beginning of the financial ruin of the Szilard family. Leo was allowed to continue engineering school until September 1917, when he was called to enter the Austro-Hungarian army. Assigned to the Reserve Officer School in Budapest, he was able to visit his family and report on military studies and life in the barracks. He amused them with stories of friendships and parties with other recruits and of the laughable idiocy of some of the officers in charge. Leo stayed in Budapest for about eight months, quite peacefully mixing barrack life with studies and visits to the Technical University and to his family.

In May 1918 Szilard was transferred from Budapest to an artillery camp in the Austrian Tyrol as an ordnance cadet assigned to study explosives and receive training in artillery and saber charging. Although World War I was raging all over Europe, these trainees had little to do other than enjoy hikes in the beautiful Tyrolean Alps. However, Szilard suddenly became very ill in late September and obtained leave to go back to Budapest, where he was admitted into an army hospital and diagnosed with the Spanish flu. By October he had reported his illness to his commanding officer, asking to be excused from military school, when he found out that his unit had been sent to the Italian front and that the Spanish flu had essentially closed the school. Only a few days before the armistice, in early November, he learned that his regiment had been attacked near Trieste, where all his comrades were lost. In a typical Szilardian twist, it appears that Leo's life was saved by the flu!

In spite of his relatively safe stint in the army, Leo felt that his exposure to the military affected him for the rest of his existence. His previous life at the Vidor Villa had been so sheltered and comfortable that simply moving with the troops and living in barracks created a feeling of insecurity. Years later he claimed that it was his experiences during the war that were responsible for his inclination to accumulate only what could fit into two suitcases so that he could move quickly in case of an emergency. It is true that the war's end marked the beginning of great turmoil throughout Europe and greater dangers for Leo. After the Russian Revolution, Czechoslovakia declared its independence and both Poland and Germany became a republic. As for Hungary, it was the end of the Austro-Hungarian Empire, and German Austria proclaimed itself a republic. The war also destroyed the Hungarian economy, as the currency was devalued and government bonds became worthless. Louis Szilard had lost his investments and his income; his only remaining asset was his elegant apartment in the villa. However, there was no coal to heat it and no money for shopping, and the last of the Szilard's savings was to pay for the children's education.

After Leo recovered from the flu, both he and Bela again joined the Budapest Technical University, but Leo was more interested in the economic situation and he followed politics closely. As the postwar Hungarian government was weakening, the communist government of Béla Kun came to power in the spring of 1919. Kun and many communist political leaders were Jewish, bringing to politics an intellectual tradition that interested Leo, who made contacts with some of the new ministers. However, after Kun's Red Army invaded Slovakia and created terror, Szilard rightly guessed that Kun's defeat was imminent and that a conservative and anti-Semitic reaction would follow. It was during the summer of 1919 that Szilard decided to leave Budapest and continue his studies in Berlin. He took multiple examinations at the university to obtain his degree, applied to the Technical Institute in Berlin, and used some of his contacts in the Kun government to obtain a passport and visa. As the Romanian troops prepared to invade Budapest under the leadership of Miklós Horthy, Leo applied to a Reform Church to be registered as a Calvinist.[9] Unfortunately, this did not prevent Leo and Bela from being beaten by anti-Semitic students when they returned to the engineering school in September.

Leo's trip from Budapest to Berlin over the Christmas holidays of 1919 was a nightmare. After the Horthy regime refused him an exit visa to study in Germany, connections in the government and bribes provided Leo with a travel pass to Berlin valid only over the holidays, from December 25 until January 5. Carrying a single suitcase and the last pound sterling notes left from his parents' fortune, Leo embarked on a daily excursion boat to Vienna on Christmas day. This sightseeing cruise on the Danube, on which he left Margaret Island and Budapest behind, was the most depressing but the easiest part of Leo's trek to freedom. In Vienna Leo obtained a student visa to enter Germany, and on New Year's Day a train took him across the border. However, coal shortages turned this one-day trip into a one-week journey as the train stopped for hours in freezing cold with little to eat. On January 6, 1920, he finally arrived in Berlin, where Bela joined him on March 1. Berlin was a crowded and busy city: housing was hard to find, food was expensive, and foreign students were unwelcome as political and social troubles were brewing all over Germany in preparation for the terror of the second World War. On January 24, in Munich, Adolf Hitler spelled out the platform of a new political party, the National-Socialist Democratic Workers' Party, later known as the Nazi Party.

The Szilard brothers enrolled in the Technical Institute of Berlin, renting a small one-bedroom sublet nearby from a motherly lady. As before, at the Budapest Technical University, Bela worked hard, and he earned a little money by tutoring students and holding a part-time job as a draftsman. Leo spent most of his time thinking; he did not seem to study or do the exercises important to engineering, except for attempts at the required drawing assignments that he left to Bela to finish for him. He was quickly losing interest in engineering and started to fantasize about physics.

In the 1920s Berlin was the mecca for modern physics and chemistry. The founder of modern (quantum) physics, Max Planck, who had received the Nobel Prize in Physics in 1918, was teaching at Berlin University. Also at the university were Max von Laue, a student of Planck and 1914 Nobel laureate, Fritz Haber, who had received the Nobel Prize in Chemistry in 1918, and Walther Nernst, who would win the Nobel in Chemistry in 1920. James Franck, a physicist from Göttingen, who was then a visitor at the university, would share the physics Nobel Prize with Gustav Hertz in 1925. Most importantly for Leo's future, Albert Einstein was also in Berlin, or rather in the suburb Dahlem, where he was director of the Kaiser Wilhelm Institute (KWI) for Physics. Einstein, who would receive the Nobel in Physics in 1921, also gave a weekly seminar and attended colloquia at the university in downtown Berlin.

Szilard was excited by his new intellectual environment and took a heavy class load. He attended lectures at the university, but the high point of his week was the physics colloquium held every Thursday afternoon. There, in the front row, sat the celebrities: von Laue, Nernst, Planck, and Einstein. Leo enjoyed not only the formal presentations but also the lively discussions and arguments and the chance to meet the professors over coffee and cake after the colloquium. At first Leo sat with the other students in the back row, but a few months later he had moved to the middle of the classroom, and soon he was sitting in the front row. This would be his regular place in seminars for the rest of his life, where he would interrupt the speaker constantly if he was interested, and fall asleep or walk out if he wasn't.

Szilard especially enjoyed the lectures and discussions with von Laue. Szilard asked von Laue to be his advisor for a doctoral thesis, and von Laue suggested a topic having to do with the theory of relativity. Szilard, however, was also interested in statistical mechanics and asked Einstein to teach a course in that field to a few friends during the fall semester in 1921. The friends Leo invited to attend included Eugene Wigner and John von Neumann, both Hungarians born in Budapest. That course was a turning point for Leo. Not only did it reinforce his friendship with Einstein, but it also made him realize that he was not very good in mathematics and that he had lost interest in his thesis topic. He had become involved with thermodynamics, and over the Christmas holidays he produced a paper that he hesitated to show to his advisor because, although von Laue was an expert in thermodynamics, it was not on the subject that he had proposed for Leo's thesis. Instead Szilard described his work to Einstein, asking for feedback and advice. With Einstein's encouragement and approval, Szilard then took his paper to von Laue, who accepted it as the thesis for his doctorate in physics, which he received in August 1922. Interestingly, throughout his doctoral work Leo also took courses in philosophy and ethics.

Having obtained his doctorate, Leo was unsure what to do next. He first thought of teaching physics, and he asked his good friend Michael Polanyi to write a letter of recommendation to a professor at the University of Frankfurt. However, Szilard really wanted to stay in Berlin and did not follow up. He had first met Michael Polanyi and his brother Karl in Budapest during the political turmoil following the war. Michael was a medical doctor who later obtained a doctorate in chemistry. In 1920 he had moved to Berlin to take an appointment at the Kaiser Wilhelm Institute for Fiber Chemistry in the Dahlem suburb of Berlin.

By 1923 he was directing a research group in the KWI for Physical Chemistry and Electrochemistry directed by Fritz Haber. At the time, Dahlem-Berlin was a major research center where several prestigious institutes were located, all under the umbrella of the Kaiser Wilhelm Society for the Advancement of Science and under the leadership of famous directors. Einstein, for example, was director of the KWI for Physics from 1917 to 1933 (Max von Laue, however, became deputy director in 1922, relieving Einstein from administrative duties). Equally famous was the KWI for Chemistry, which included a radioactivity department headed by Otto Hahn and Lise Meitner, who in 1939 would discover nuclear fission.

The theologian Adolf von Harnack was the first president of the Kaiser Wilhelm Society, which he founded in Berlin in 1911 to promote science in Germany. Max Planck succeeded Harnack in the office from 1930 until 1937. Harnack had introduced the then-original concept of establishing a research institute outside a university. This was achieved by founding institutions independent of the state and its administration. Money was raised within Germany primarily from individuals, industry, and scientific organizations. Contributions from abroad included funds from the Rockefeller Foundation. The Kaiser Wilhelm institutes were well funded and focused on excellence in research, and their formal and financial independence from the universities relieved their participants from teaching obligations. Since the Kaiser Wilhelm institutes had been established to develop Germany's science and technology, their research was expected to have practical applications, in contrast to the more theoretical research carried out in universities.[10]

Many KWI scientists, including Einstein, von Laue, and Polanyi, commuted on the subway between Dahlem and downtown Berlin to attend seminars and colloquia. So did Szilard's friend Eugene Wigner, who was conducting doctoral thesis experiments at the KWI under the direction of Michael Polanyi. Polanyi had introduced Leo to the chemist Hermann Mark, with whom he started a collaboration on X-ray diffraction studies. Soon Szilard became a fixture at KWI, although he had no position except as a visiting researcher receiving a small consultant fee. It is unknown how he managed to survive after his parents' money ran out. He mentioned that he was hungry during his doctorate work; later he often had dinner at the homes of friends or acquaintances who enjoyed his lively spirit and wry humor. At KWI he seemed to be loitering, asking questions, criticizing others, and telling people what experiments to do. This activity, as an uninvolved catalyst, was to be his style for life, just as he directed the games of his playmates at the Vidor Villa. Some resented his arrogance and the fact that he patented all his ideas. He filed his first patent in 1923, for an X-ray sensor that he conceived on the basis of Hermann Mark's equipment. Although today it is commonplace for scientists to patent their ideas, at the time it was viewed as unscientific and mercenary. Without a real job, however, Leo had to hope that the patents might provide him with enough income to live without being pinned down by a regular job. Meanwhile, he was networking, intruding on conversations, and scrounging around for food and facts. He owned only some worn-out clothes and a few books and was constantly moving from one rented room to another. Not yet a roving scientist, at this point he was more of an intellectual vagabond. This would become a lifelong trend of his career: an academic position unclear, a material situation precarious, and restlessness.

Eventually, in 1925, he began teaching as von Laue's assistant—his first true academic position. In 1926 Szilard and Einstein began working together on a very practical application of physics: a refrigerator pump with no moving parts. By 1927 Szilard and Einstein had filed the first of eight patents for their electromagnetic pump and parts. Einstein was comfortable with patents; in fact, he had been working as a clerk in a patent office at the time he developed his theory of relativity. In 1928 Szilard patented the concept of a linear accelerator, in 1929 a cyclotron, and in 1931 the principle of an electron microscope. Although he enjoyed playing with ideas, Szilard rarely followed through, and by 1930, after ten years as a physicist, he had few substantial achievements. His most important encounter during that period in his life was one that he had hardly noticed: he met Gertrud (Trude) Weiss, an exceptional Viennese woman who would become his wife in 1951.

Leo and Trude met in Berlin in the fall of 1929 through their mutual friends Karl and Michael Polanyi. Trude had enrolled at the university in physics and biology classes and sat in for a few weeks on Leo's classes entitled "New Questions in Theoretical Physics." She asked questions, they started talking, and they attended some parties together.[11] When Leo inquired about her plans after the courses ended, she told him that she was undecided between physics and medicine. In his typical blunt style he told her that she was not smart enough for physics and advised her to return home to Vienna and study medicine. Then twenty years old, the daughter of a successful physician, Trude returned to Vienna, moved into the comfortable apartment of her parents, and enrolled as a medical student. This is where Leo called on her in 1933, four years after their first encounter in Berlin, which had been followed by exchanges of friendly but formal letters.

Much had happened in those four years. During a deep worldwide economic depression, Hitler had come to power in January 1933. At the very same time, the newly elected American president, Franklin D. Roosevelt, was assembling his cabinet. By 1933 most of Szilard's friends had left Berlin. Both Einstein and von Neumann were among the first permanent faculty members of the Institute for Advanced Study at Princeton. Wigner had moved to Princeton University, and Michael Polyani had accepted a position at the University of Manchester. Wigner and Polanyi worried about their old friend's future, especially after Wigner received, in October 1932, a letter revealing Szilard's internal conflicts and doubts about pursuing science.

For Szilard, there were causes that were nobler and more important than science. Unfortunately, no institution would provide the leisure time and freedom necessary to address such issues. He had failed to become financially independent with his patents and considered conventional teaching and research positions unsuitable for his purposes. Wigner and Polanyi had an idea: get Einstein to recommend Szilard for a job at the Institute for Advanced Study at Princeton. But Einstein considered Szilard an experimental rather than a mathematical physicist and declined to recommend him.

On March 30, 1933, Szilard packed his two bags of belongings and embarked on an express train out of Germany to Vienna. Joseph Goebbels, the Reich propaganda minister, had announced an anti-Jewish boycott to start the next day. On April 1, the train leaving Berlin would be overcrowded and Hitler's storm troopers would question and seize the valuables of passengers that they suspected to be non-Aryan. In Vienna, Szilard met with his brother, Trude, and many old friends. However, his major preoccupation was the organization of a system to relocate academic refugees, both faculty and students who had been dismissed from universities in Nazi Germany. During the spring and summer of 1933 Szilard made numerous trips to London for that purpose, where he helped to organize the Academic Assistance Council.

Szilard felt at ease in England, and he liked the English reserve and humor. He was even considering a teaching position at the University of London when, in September, he read in the *Times* about a lecture given by Lord Ernest Rutherford at the annual meeting of the British Association for the Advancement of Science. Rutherford, a highly respected nuclear physicist, was then director of the famous Cavendish Laboratory at Cambridge. In his talk he had summarized the discoveries of the last quarter century on atomic transmutation and concluded, "Anyone who looked for a source of power in the transformation of the atoms was talking moonshine."[12] This bold statement by a famous expert immediately provoked Szilard's contrarian spirit, and he dreamed of proving the expert wrong. It is while pondering that statement during a stroll through London that he conceived the mechanism that could release the enormous energy locked into an atom: a nuclear chain reaction and the critical mass that would start and sustain it. He quickly understood the possible consequences, from generating beneficial energy to building destructive atomic bombs.

This revelation rekindled Szilard's enthusiasm for research in the field of nuclear physics. By March 1934 he had filed his first British patent application, "Transmutation of Chemical Elements." After several improvements the patent was accepted two years later, but, assigned to the British Admiralty, it was kept secret and withheld from publication until 1949. Szilard was eager to test his chain-reaction ideas, but he feared that the Germans might hear about them. This made him cautious in describing his ideas to colleagues and leery about working in an academic laboratory. While attempting to interest financiers in his chain-reaction research he remained vague, promising possible applications that seemed impractical, which left them unconvinced. Without a position providing him with his own laboratory, he worked as a summer guest with Chalmers in London, and for a few months he worked as a research associate at New York University. Finally, in 1935 he accepted a research fellowship at the Clarendon Laboratory in Oxford, which allowed him to spend half of his time working in the United States. He was in New York in September 1938 when the Munich Agreement was reached permitting Germany to annex part of Czechoslovakia. Fearing war in Europe, he decided to remain in America and resigned from Oxford.

It was during January 1939, on a visit to his old friend Wigner in Princeton, that Szilard heard the stunning news: the fission of uranium had been demonstrated! The element capable of carrying out the process that he had postulated in 1933 had been found. Not surprisingly, the discovery had been made at the KWI for Chemistry of Berlin-Dahlem. Niels Bohr had heard the news from Lise Meitner, who, because she was Jewish, had fled Nazi Germany for Sweden in 1938. She had remained in touch with Otto Hahn, her colleague at KWI who had written her about an experimental result he did not understand: bombarding uranium with neutrons yielded a lighter element, barium. She and her nephew Otto Frisch, a physicist who was working in Bohr's lab in Copenhagen, published a paper in February 1939 explaining the phenomenon, estimating the energy it released and giving it the name "fission."[13]

Szilard knew it was essential to establish whether neutrons were emitted in the process. Szilard, jobless as usual, had to borrow money and ask permission of Columbia University to work as a guest. A clumsy experimentalist himself, he also had to find a skillful collaborator, and he convinced Walter Zinn, a Canadian physicist, to join him. They discovered that neutrons were, indeed, emitted upon uranium fission, making the chain reaction an ominous possibility. Enrico Fermi, Frédéric Joliot, and their coworkers made the same observation at about the same time. In the spring of 1939 Szilard and Fermi did further work in collaboration, and, in spite of a difficult relationship, they published a paper together in August.[14]

Szilard was too familiar with the Berlin-Dahlem KWI research center and with the concentration of talent and technology in Germany to underestimate the country's power. There was no doubt in his mind that the race was on, and that Nazi Germany was working on developing an atomic bomb.[15] This is when, in the summer of 1939, Szilard decided to contact his mentor Einstein to help warn the U.S. government by sending the famous letter to President Roosevelt (see Prologue). In 1940 Columbia University was granted a government contract to develop the chain reaction system, and Szilard became a member of Columbia's National Defense Research Staff. By 1942 the group working on the bomb, including Szilard and Fermi, was moved to the University of Chicago as the Metallurgical Laboratory, a code name for part of the Manhattan Project.[16]

As a member of a team, Szilard could be a great catalyst. On his own, however, he was too impatient and awkward to be a good experimental physicist, and he was not good enough at mathematics to be a mathematical physicist. However, his interactions with team members could be rocky because he was blunt, bossy, and constantly contrary. By the time the Manhattan Project ended Szilard had acquired a widespread reputation as a troublemaker. All this did not make it easy for him to find a job as a physicist after the war. During the Manhattan Project he had antagonized the government, and the military in particular, through his clashes with the brass. He had no chance for a job with the new Atomic Energy Commission, the organization that took over control of atomic energy from the army. His 1945 petition not to drop the bomb on Japanese cities had even angered Oppie in the Los Alamos days. Szilard could forget about ever getting a job at the Institute for Advanced Study at Princeton, where Oppenheimer had become director in 1947. His relationship with Fermi had been so painful during their collaboration that a job at the University of Chicago's new Institute of Physics, directed by Fermi at the time, was out of the question as well. To find a job after the war, Szilard needed a paradigm shift. This is when he turned to biology.

In the years following World War II biology was relatively unsophisticated compared to physics, and the experimentation appeared simpler. The field of biology raised social and political issues that interested Szilard at least as much as the science itself. Moreover, the barriers between the scientific disciplines were breaking down at the time: biochemistry and biophysics appeared as the field of molecular biology emerged.[17] Many of the brightest physicists of the day were turning to biology, which was appealing because it had not yet undergone the conceptual revolution that had changed physics in the 1920s and 1930s. One could predict that, with all the technical advances that had taken place in the early 1940s, a revolution would soon change biology as well. After the war, some physicists wanted to become involved in that revolution, whether they were inspired by curiosity or guilt about the A-bomb or were just looking for a job in an emerging field. By setting up extraordinary circumstances, World War II played an essential role in the development of the new discipline of molecular biology in the United States.

CHAPTER 4

Atoms in Biology

... dreams of doing simple experiments on something like
atoms in biology.

—Max Delbrück[1]

Max Delbrück epitomizes the physicist-turned-biologist, and, by attracting other outsiders to this emerging field, he became a major founder of molecular biology. In 1906 he was born into a distinguished family in Berlin: the famous chemist Justus von Liebig was his mother's grandfather, and his uncle was Adolf von Harnack, the founding president of the Kaiser Wilhelm Society. After completing his graduate studies in theoretical physics in Göttingen in 1929, Delbrück took a job for about a year at the University of Bristol. This was his first time outside Germany and his first exposure to a different culture and language. He then obtained a Rockefeller Fellowship to do postdoctoral work in Copenhagen and Zurich. It was Niels Bohr, in Copenhagen, who had the most influence on Delbrück's future by getting him interested in biology. By 1932 two jobs were open to him: he could become an assistant to either Lise Meitner in Berlin or Wolfgang Pauli in Zurich. He chose the job in Berlin because the Kaiser Wilhelm Institute (KWI) for Biology was located close to the KWI for Chemistry at Berlin-Dahlem, where Meitner was codirector with Otto Hahn. He arrived in Berlin in the fall of 1932 and stayed for five years.

Delbrück became an advisor to Meitner on theoretical physics at a time when Meitner and Hahn were irradiating various elements, among them uranium, with neutrons. Delbrück was supposed to interpret the results of these experiments, but he entirely missed what was really going on: fission. He was more interested in private discussions taking place mostly at his mother's house. There, a small group of physicists joined by a few biologists and biochemists held private seminars. Among them was Nikolaï Timoféeff-Ressovsky, a geneticist from the Kaiser Wilhelm Brain Research Institute in Berlin-Buch, a suburb quite distant from Berlin-Dahlem. The American geneticist Hermann J. Muller had spent one year in Timoféeff's lab in 1932. Following up on Muller's discovery of radiation-induced mutations, Timoféeff was studying the induction of mutations in *Drosophila* by ionizing radiation. This was an attractive way to bridge physics and genetics, and those informal seminars resulted in a collaboration between Delbrück, Timoféeff, and the physical chemist K.G. Zimmer. Together in 1935 they published "On the Nature of Gene Mutation and Gene Structure" in an obscure journal.[2] However, many reprints were mailed out, and that paper was well received by both physicists and biologists. In 1936 Delbrück was invited to apply for a second Rockefeller Fellowship. He chose to work at the California Institute of Technology (Caltech) in Pasadena, California, where he arrived in 1937.

By then Delbrück was disenchanted with the complexity of *Drosophila* and intrigued by the crystallization of the tobacco mosaic virus achieved in 1935 by Wendell Stanley at the Rockefeller Institute. He was determined to turn to virus research and discovered at Caltech the simplicity of working with viruses that attack bacteria, bacteriophages (phages for short). There Emory Ellis was using an easy assay for phages: each phage particle could be visualized as a hole or plaque in a lawn of bacteria after overnight incubation. The plaque method had been described in 1917 by Félix d'Herelle, who coined the name "bacteriophage" for bacterial viruses.[3] These plaques could be seen with the naked eye, counted, and examined, and the conditions of phage replication could be controlled. This meant that the essential property of life, replication, could now be studied simply and quantitatively. Delbrück saw phages as atoms in biology!

By September 1939 his Rockefeller Fellowship was running out, but Delbrück decided to remain in the United States. Although he was not Jewish, he was concerned about developing his career in Nazi Germany. Thomas Hunt Morgan, the head of Caltech's biology division, recommended that his fellowship be extended. With the help of the Rockefeller Foundation, Delbrück obtained a lectureship in the Physics Department of Vanderbilt University in Nashville, Tennessee. There 50 percent of his salary would be supported by the Rockefeller Foundation, which would allow him to pursue his research in biology.

Delbrück left Pasadena a few days after Christmas 1939 and arrived in Nashville on the eve of the New Year (1940).

Meanwhile, far away in Paris, France, Salvadore Luria was working at the Curie Laboratory of the Institut du radium on the effect of radiations on phages.[4] Born in 1912 in Turin, Italy, into a middle-class Jewish family, he had graduated from the University of Turin medical school in 1935, more to please his parents than because he was attracted to medicine. His high school friend Ugo Fano had chosen physics and had encouraged Luria to become a scientist. This was in the late 1930s, and Ugo was telling extraordinary stories about new ideas in physics and about his heroes Bohr, Schrödinger, Heisenberg, and, above all, the Italian physicist Enrico Fermi, who was then working in Rome. While in medical school Luria tasted research when he worked in the laboratory of the well-known histologist Giuseppe Levi, but histology was not for him. He tried to find a medical specialty that could bridge physics and medicine and joined the radiology department. This, however, he found to be the dullest of specialties.

After a stint in the army while Ugo was working with Fermi, in 1937 Luria moved to Rome to finish his radiology course and learn some physics. The year he spent among physicists introduced him to radiation biology and was crucial in other ways as well. There he met not only Enrico Fermi but also Franco Rasetti, a professor with encyclopedic knowledge who was teaching spectroscopy.[5] Rasetti introduced Luria to articles by Delbrück and to the concept of genes as molecules. One day in 1938, while riding a tram in Rome thinking about Delbrück's papers, he met a professor of virology named Geo Rita. It turned out that Professor Rita was analyzing water from the Tiber River that contained bacteriophages. Luria had never heard of phages but spent some time in the professor's lab growing and counting them. He soon realized that phages could help in the investigation of Delbrück's idea about genes. It is amazing that Delbrück and Luria rediscovered phages around the same time, independently, and thousands of miles apart. When Luria saw the first article by Delbrück on phages, he knew that they were thinking along the same lines and started dreaming of joining him in Pasadena.

Political upheaval delayed their meeting. On July 17, 1938, Mussolini issued the Racial Manifesto, officially siding with the anti-Semitic policy of Nazi Germany. This is when Luria left Italy for France. His Jewish friend Ugo would soon emigrate to the United States, and Fermi, whose wife was Jewish, had left Rome for New York with a stopover in

Stockholm to accept the Nobel Prize. The awarding of the prize to Fermi helped the couple to escape from Fascist Italy, making this the most welcome and useful Nobel Prize ever awarded.[6] Luria stayed in Paris until June 1940, when, after invading Belgium, German troops had entered France. As Paris emptied, Luria traveled by bicycle and freight train to Marseilles, the site of multiple consulates, where he eventually obtained the necessary visas to exit France, transit through Spain and Portugal, and enter the United States. In Lisbon he boarded the Greek ship the SS *Nea Hellas* to New York, and he sighted Lady Liberty ten days later, on September 12, 1940.

In New York Luria visited Fermi at Columbia, and with his help he obtained fellowships and a small laboratory space at the College of Physicians and Surgeons. He applied for American citizenship, at the same time changing his name from *Salvadore* to *Salvador E.,* the letter *E* now standing for *Edward.* His new American friends, which he quickly made, all called him Salva. Finally, two days before the beginning of 1941, Salva met Max. They had arranged by mail to meet in Philadelphia at the meeting of the American Physical Society. They spent January making plans and playing with phages in Luria's little lab at Columbia.

Delbrück and Luria decided to spend time working together at the Cold Spring Harbor Biological Laboratories, on Long Island, during the summer of 1941; this was to be the first of many summers they spent there. That summer of 1941 also reunited Salva with his old friend Ugo Fano, brought to Cold Spring Harbor by Milislav Demerec, the new director of the Biological Laboratories. Since 1933 Cold Spring Harbor had been the site of the annual Cold Spring Harbor Symposium on Quantitative Biology. Since their inception these symposia had emphasized the importance of incorporating the approaches of chemistry, physics, and mathematics into biological research. Demerec made these symposia, and the sale of the symposia red books, a remarkable success. Moreover, by extending to Delbrück and Luria the hospitality of Cold Spring Harbor labs over the summers, Demerec made Cold Spring Harbor a meeting and training center for molecular geneticists.[7]

In December 1941, while Luria was in Philadelphia taking the first electron micrographs of phages in collaboration with Tom Anderson, the Japanese attacked Pearl Harbor. As the United States entered the war, both Salva and Max suddenly became enemy aliens. They had to register with the U.S. government but were not otherwise troubled.[8] Actually, this was a quiet time for them to work, as they had few students and visitors or scientific meetings to attend; even the Cold Spring Harbor symposia were cancelled for three years. During this period Luria's beautiful phage electron micrographs led to a publication and a Guggenheim fellowship that allowed him to join Delbrück at Vanderbilt. However, after less than a year in Nashville Salva moved to Bloomington, Indiana, the seat of the state university. This move was not entirely his choice, as his 1940 Rockefeller Foundation Fellowship had stipulated that he was required to accept the first respectable academic position offered to him, and Indiana University's offer was respectable enough.

Bloomington was not New York City, but Indiana University had a pleasant campus and attracted some leading geneticists. Luria remained there for seven productive years, after which he moved to the University of Illinois in Urbana. It was soon after his arrival in Bloomington that Luria conceived and performed the famous fluctuation test experiments that demonstrated that mutations occurred spontaneously in bacteria. He discussed his idea with Delbrück, who immediately figured out the mathematical theory, and the two published a landmark paper in 1943.[9] These experiments also made it possible to calculate mutation rates and started the field of bacterial genetics. Moreover, it was soon established that mutations also occur in phages. As it turned out, bacterial and phage genes not only replicate and mutate but also recombine. Because of their rapid replication and small size, which allows the observation of very large populations, bacteria and their viruses became the ideal tools for genetic research, and they took over the field of genetics in the 1950s and 1960s. Microbial genetics became essential in the development of molecular biology, a new approach combining genetics with biochemistry to characterize life at the molecular level. The revolution in biology was on the move, as molecular biology led to recombinant DNA, which gave rise to today's biotechnology. And then there was the mind-boggling possibility that "what is true for bacteria is true for the elephant." Although this is something of an overstatement, it is true that the basic processes and molecules of life are largely conserved between organisms.

Alfred Hershey, who had been doing some work on phages in the Bacteriology Department of Washington University Medical School in St. Louis, Missouri, visited Delbrück in Nashville and Luria in Bloomington. He immediately joined them to form the nucleus of the famous Phage Group. The American Midwest became the cradle of molecular genetics, and the trio—Delbrück, Hershey, and Luria—later shared a

Nobel Prize in 1969 for their discoveries concerning the replication mechanism and genetic structure of viruses. By 1946, World War II had ended and the Cold Spring Harbor symposia resumed. The first postwar Cold Spring Harbor symposium, entitled "Heredity and Variation in Microorganisms," attracted a remarkable group of speakers and participants. That symposium was seminal not only because of the importance of the scientific presentations but also because the participants started a network of scientists that would shape the future of biology and influence the creation of the Salk Institute. At that first postwar symposium at Cold Spring Harbor, lifetime friendships were formed, future sabbatical visits were planned, and collaborations and exchanges of students were initiated (see frontispiece).

By 1945 Delbrück had organized at Cold Spring Harbor a summer phage course, which he taught himself for the first three years. Mark Adams, from the Microbiology Department of New York University, had taken the course the second year and taught it every year starting in 1948.[10] This course recruited an entire generation of new phage researchers.[11] It also introduced to molecular biology many scientists working in other fields, including physicists considering a switch to biology. One of them was Leo Szilard, who took the phage course in 1947.

After the war, in October 1946, Szilard had become professor of biophysics at the new Institute of Radiobiology and Biophysics at the University of Chicago. As usual, his academic position was complicated: he was half-time professor of biophysics and half-time advisor to a project in the Division of Social Sciences inquiring into the social aspects of atomic energy. That tailor-made position, arranged by Robert Hutchins, the university president and a supporter of Szilard, freed him from the usual requirements of a university professorship such as teaching and research. It allowed Szilard to travel, learn some biology, and get involved in politics. He best described his unusual academic job in a 1950 letter to Niels Bohr: "Theoretically I am supposed to divide my time between finding what life is and trying to preserve it by saving the world. At present the world seems to be beyond saving, and that leaves me more time free for biology."[12] He received a quick but superficial education in modern biology by taking specialized courses, such as Delbrück's phage course at Cold Spring Harbor and Cornelius Van Niel's microbiology course in Pacific Grove, and attending numerous biology meetings. He was mostly learning by networking with the founders of molecular biology.

As for biological research, he would not consider approaching experimental research without a collaborator to carry out most of the work. In January 1948 he teamed up with Aaron Novick, a physical chemist he had met during the uranium project. Together they built a small lab in the basement of a synagogue that belonged to the University of Chicago. They collaborated on half a dozen papers on bacterial mutations,[13] and they worked together until 1953, when Novick left Chicago to work at the Pasteur Institute in Paris. Szilard then closed his biological lab, and by 1954 the Institute of Radiobiology and Biophysics at Chicago had also dissolved. He then became, somewhat to his embarrassment, a full-time professor of social sciences. Szilard remained involved with physics, however, by consulting with industrial firms. Since his Berlin days with Einstein he had been convinced that there should be no division between applied and basic science, and that both approaches could benefit from and should interact with each other.

Meanwhile Gertrud Weiss was teaching at the University of Colorado Department of Preventive Medicine, where she had been on the faculty since 1950. In total secrecy, Leo and Trude were married in 1951. Whether he wanted to be with Trude or to escape his uncomfortable position in Chicago or both, Szilard tried to obtain a position in Denver. However, Ted Puck, the biophysicist at the University of Colorado Medical Center, turned Szilard down. He did it in the nicest way possible in a touching handwritten letter.[14] Puck explained to Szilard, "Your mind is so much more powerful than mine that I find it impossible when I am with you to resist the tremendous polarizing forces of your ideas and outlook." In a very constructive fashion, Puck then suggested that Szilard apply for a roving commission. That would allow him to divide his time between four or five laboratories and provide consulting and stimulation to the other scientists. Actually, that was precisely what Szilard wanted: to become a roving theoretical biologist. He applied for a grant from the National Science Foundation (NSF), but his grant was turned down in 1956, leaving his academic future uncertain once more. This is when he and Jonas Salk first met, on October 25, 1956.

Szilard admired Salk for driving the polio problem all the way from basic research—establishing the number of types of poliovirus—to the development of the vaccine itself. Both men were interested in creating an institute devoted not only to studying public health problems but also to finding practical solutions. We will never know who was first to think about creating such an institute, but the idea most likely resulted from the first interaction between Salk and Szilard. What is clear is that soon after that first meeting each of them started working—independently, and in his own way—to realize the idea. Salk became actively involved in attempting to create such an institute on the campus of the University of Pittsburgh (see chapter 2). As for Szilard, he started contacting influential people and putting his ideas on paper with the help of his old friend Bill Doering.

Szilard had met William von Eggers Doering in 1952 while they both served as consultants for the Conservation Foundation, a temporary committee set up during President Truman's administration to explore issues related to the conservation of natural resources. Szilard must have convinced someone in the government of the obvious: that population control was a sure way to conserve resources. In any case, he obtained a grant to invite scientists to New York to determine the state of knowledge in the area of human reproduction, with the goal of identifying novel contraceptive agents. Szilard was obstinately opposed to mechanical vaginal inserts. Thinking mostly of rice-consuming developing countries, he favored the idea of introducing antifertility rice, which would be sold at a lower price than regular rice to create an incentive for buying it. He decided that he needed on this committee not only biochemists but an organic chemist as well, and someone recommended Doering. The most positive result of those meetings was to introduce the two men. They enjoyed their collaboration and remained friends until Szilard's death.

Doering was then professor of organic chemistry at Yale. He had become well known in 1944, when, at age twenty-seven, he reported the total synthesis of quinine.[15] Synthesis of quinine was timely in the 1940s, as that drug was badly needed by soldiers fighting in the Pacific at the same time that the war had interrupted the supply of the natural compound.[16] The synthesis had been done in collaboration with Robert B. Woodward, who won the 1965 Nobel Prize in Chemistry for his outstanding achievements in the art of organic synthesis. It also launched the brilliant career of Doering, who eventually moved to Harvard, where he remained active as an emeritus professor until his death in 2011.[17] A by-product of the contribution of Doering and Szilard to the work of the Conservation Foundation was a lengthy memorandum written by the two proposing the creation of interacting research institutes in the area of public health.[18] That memo, dated January 4, 1957, was addressed a few days later to Cass Canfield, a powerful voice and a powerful man who was the publisher and senior editor of Harper & Brothers.

Canfield's major interest outside publishing was birth control. As chairman of the executive committee of the Planned Parenthood Federation, he traveled, raised funds, and spoke in favor of the cause. He had close connections with John Cowles Sr., another giant in the publishing world who, in 1950, was appointed director and later trustee of the Ford Foundation. Moreover, Cowles's wife, Elizabeth, was actively involved with a number of civic organizations and in 1934 had founded the Iowa Maternal Health League, which later became part of Planned Parenthood.

After his first meeting with Salk, in October 1956, Szilard called upon his connections with rich and powerful people involved with Planned Parenthood. He got Canfield to set up a meeting with Cowles and his wife. That meeting took place barely three weeks after Szilard's first meeting with Salk.[19] After the meeting with the Cowles, Canfield and his wife, Jane Fuller Canfield, had dinner with Szilard and Doering to discuss procedure and possible trustees of the planned institute. It was decided that this group should include top scientists to broaden the research interests of the new institute. At that dinner Szilard and Doering agreed to outline their plans, extending the interest of the proposed institute beyond birth control to various areas related to public health. The Doering-Szilard memorandum was the result of those discussions.[20] By the time Szilard met Salk again, in January 1957, that memorandum was ready.

The memorandum was entitled "A proposal to create two independent research institutes operating in the general area of public health, designated as: Research Institute for Fundamental Biology and Public Health, and Institute for Problem Studies." The Doering-Szilard memo was sent not only to Salk but also to other prominent scientists, including the biochemist Fritz Lipmann, the chemist Linus Pauling, and the geneticist Hermann Muller. It proposed to create a basic biological research institute where no teaching would be required. Instead of teaching, its members would get involved in projects, developed in a second institute, that aimed to find practical solutions to health problems. This scheme was clearly inspired by the Kaiser Wilhelm institutes, which Szilard had observed in Berlin. There the basic research was carried out at the university, while those working at the Kaiser Wilhelm institutes were expected to produce practical applications instead of teach. In answer to the question posed by Linus Pauling, "Why two institutes?" Szilard explained that the source of money to run each institute would be different. This was also the main reason to separate the Kaiser Wilhelm institutes from the university in Berlin.[21]

The areas of research defined in the memo focused on public health, most likely inspired by Trude; they included birth control, cigarette smoking, and coronary disease. In addition, Szilard proposed that the Institute for Problems Studies expand into what he called "political thought," reflecting Szilard's interests in politics. This would involve bringing in, perhaps as visiting members, scholars in areas outside science such as historians and economists. Szilard then described staffing that would include, in addition to the regular staff members, an advisory body of affiliate members.

In a cover letter Szilard invited Salk to consider a position as affiliate member of the Research Institute.[22] He also mentioned that Canfield was prepared to explore possible funding sources, while Cowles had already discussed those plans with the incumbent president of the Ford Foundation, Henry Heald, and with his predecessor, Rowan Gaither. Szilard met Salk again one week later, on January 26, in Chicago, where they further discussed ideas for an institute.[23] Salk made up his mind almost immediately after his return from Chicago. He sent his negative answer to Canfield and Szilard on February 8.[24] In a somewhat wordy letter, Salk acknowledged that the ideas put forth by Doering and Szilard were certainly worthy. However, he raised numerous questions that indicated he considered the plans too vague, ambitious, and premature. Finally, he declined to participate in order to utilize his time and energy in an area that was already charted (i.e., the polio vaccine problem).

One should remember that a new chancellor, Edward Litchfield, had taken office at the University of Pittsburgh (Pitt) a few months earlier (see chapter 2). He had announced grand plans to turn Pitt into an outstanding university that would promote research. Salk was then at the height of his fame and had reason to expect to be included in the chancellor's project to develop research on the campus. Indeed, by May, only three months after declining to participate in Szilard's plan, Salk was informed that the Pittsburgh Municipal Hospital had been gifted to Pitt. It was to be renamed Salk Hall and provide space for his research institute on the campus. Salk was a practical dreamer and decided to play out the situation offered in Pittsburgh. There was no chance that he would pass up that solid opportunity until the fall of 1958, when the Pittsburgh option crumbled.[25]

Nothing ever came of the Doering-Szilard plan to create two interacting institutes that would cooperate on basic research projects and their applications. No such institute was realized because it was too complicated, and the funding did not become available. Because of Canfield's interest in birth control, Szilard had counted on him to raise American funds for an institute to be established in the United States. However, there was little interest in the United States in birth control methods designed for "underdeveloped" countries.[26] Szilard then conceived complicated and unrealistic schemes to solicit contributions from the Chinese, Russian, and Indian governments to endow an institute to be located in England. When these appeared impractical, he presented the funding of his institute as an opportunity for American-Russian cooperation.[27] All of these ideas were typical of Szilard: they were imaginative but naïve, and Szilard was especially unrealistic about international relations. Although he was interested in politics and well informed, he knew little about how decisions related to world affairs were made. Besides, he had no notion of or respect for the art of diplomacy, whether he was dealing with nations or with people.

However, a substantial confidential appendix to the Doering-Szilard memo remains relevant.[28] Szilard had specified in his cover letter to Salk that this appendix "contains my personal guesses on what kind of work might be done in the Institutes and by what kind of people." Six areas of research interests were considered: "I. Mammalian Reproduction; II. Protein Synthesis; III. Fat Metabolism and Coronary Disease; IV. The Effect of Cigarette Smoking on Longevity; V. Advances in Neurology—Pharmacology of the Nervous System—Mental Health and Sleep; VI. The Problem of Aging."

Public health admittedly covers many issues. Szilard had little to suggest about most of the topics, except for what he "somewhat sloppily labeled Protein Synthesis." That subject was of such great intrinsic interest that it had attracted the scientists involved in the postwar revolution in biology, the group of new biologists with whom Szilard had been networking since the mid-1940s. He wanted to attract members of that group under one roof, in a new institute that today would be called a molecular biology institute. Szilard, however, never used that term. When considering protein synthesis he went into extensive and specific detail. In the area of microbial genetics, he mentioned Joshua Lederberg and François Jacob. As tissue-culture specialists he picked Renato Dulbecco, Theodore Puck, and Louis Siminovitch.

Szilard's recommendations in the area of antibody formation were Melvin Cohn, Edwin Lennox, Howard Green, and Donald Steiner. In the field of modern microbiology he identified Seymour Benzer, Alan Garen, Norton Zinder, Maurice Fox, and Milton Weiner. It is remarkable that this list includes the three biologists who would actually join Salk to become the founders of the Salk Institute: Dulbecco, Cohn, and Lennox.

It is interesting and revealing of Szilard's personality to consider why, three years after the fact, he requested that a copy of the Doering-Szilard memo and its appendix be sent to Cohn.[29] There is little doubt that this was Leo's way of claiming credit for the idea of creating the Salk Institute. Since his youth in the Vidor Villa, Leo had enjoyed inventing complicated games in which he did not physically participate but for which he took the credit. It has often been said that Szilard was generous with his ideas. It is true that he threw them around and gave them away. His friend the physicist Isidor Rabi once told him, "You have too many ideas. Please go away."[30] Jacques Monod also remarked that Leo was "too rich in ideas . . . that he loved ideas but only as toys to be played with."[31] Leo knew that he would never realize any of them himself and always hoped that someone else would pick them up and contribute the necessary patience and hard work to realize them. When it came to the creation of the Salk Institute, Szilard was no different. He contributed the idea to make it a pioneering molecular biology institute and facilitated its location in La Jolla (see chapter 5). However, he did not participate in the hard work of organizing it, raising funds, and building it. He left that to Jonas Salk and other founders. The September 1958 letter that Salk sent to O'Connor (see chapter 3) reveals that Salk had a pretty clear opinion of Szilard's personality and motivations. Of Szilard, Salk said he "spread ideas as a bee does pollen." Salk reflected that Szilard, in a chronically precarious academic position, was interested in a new institute "so that he would have a place to visit or to have as a home base."

Both Salk and Szilard achieved breakthroughs that had profound consequences for human welfare. However, while Salk derived great fame for developing the first effective polio vaccine, Szilard remained essentially unknown for conceiving nuclear energy. This can be largely accounted for by the sources of funding for each project. The polio vaccine was developed with the support of public donations raised by the March of Dimes. Therefore, Salk's efforts and progress had to be widely publicized and he became a hero. The breakthrough of Szilard that launched the Manhattan Project was to remain secret. Moreover, no publicity was necessary because it was supported by the U.S. government using taxpayer's money. Both men, however, derived from their success the belief that they were invincible and that they were destined to save the world. Salk was to save humanity from pestilence and death, while Szilard would rid it of war and famine. Considering the scope of these tasks, they both did quite well!

CHAPTER 5

What Was It about La Jolla?

La Holla? Layola? Lahoya?[1]

By 1960 a number of people had heard about La Jolla, California, but even among those people many still did not know how to spell it. It was not that easy to find out since La Jolla was literally not yet on the map. Jonas Salk, however, had been familiar with the name since 1956. This is when a collaborator from his University of Pittsburgh laboratory, Lenora Brown, had left Pittsburgh for La Jolla with her husband, Massoud Simnad. Massoud had been offered a job he could not refuse at General Atomic, a new division of General Dynamics that had just been founded in La Jolla, a suburb of San Diego. Of Iranian origin, educated in London, and with a Ph.D. from Cambridge, he had been a faculty member at the Carnegie Institute of Technology in Pittsburgh since 1949. A brilliant metallurgist and an expert on nuclear energy and materials, he became the senior technical advisor on materials and fuels at General Atomic in 1956.

Although Massoud had an excellent job and a successful career at General Atomic, Lenora had a very hard time finding a job in La Jolla. She was an M.D. who had worked with Salk in Pittsburgh since 1953 during the polio clinical trials. She had hoped to join a virology research unit being planned at the Scripps Clinic in downtown La Jolla. She had contacted Dr. Edmund Keeney who had taken the direction of the small Scripps Metabolic Clinic in 1955 and changed its name to Scripps Clinic and Research Foundation. With a bachelor's degree from Indiana University and a medical degree from Johns Hopkins, Dr. Keeney had big dreams of creating in La Jolla a Rockefeller Institute of the West Coast. Amazingly, he succeeded beyond all expectations, although the project took a few years. In 1956, when Lenora visited him, a virus laboratory for basic research at Scripps Clinic was only an idea. She could not pin down Dr. Keeney, who made no concrete plans or commitments. She explained her difficult position in La Jolla in lengthy letters to Salk.[2] She was not very enthusiastic about setting up a lab at Scripps, but she had few options. When it came to medical research, La Jolla was a beautiful desert.

She had, however, a generally positive outlook. In her July 1956 letter to Salk, Lenora described encouraging signs of a bright future for La Jolla, including a

> possible future relationship between the Scripps [Clinic] group and the University of California's school for advanced studies in the physical and biological sciences and engineering. This is to be built here in the very near future. The land is already staked out. It is adjacent to the Oceanographic Institute and to the site of General Atomic laboratories (Massoud's group). Massoud and his new colleagues are very keen about their tie-up with the University, too. So it looks as if this area will have a good academic atmosphere along with all its other wonderful aspects.

By October 1956, still looking for a job, her letter to Salk introduced an important character in the history of La Jolla, Roger Revelle:

> The Scripps Institute of Oceanography may be a possibility. This is an integral part of the University of California, and has nothing to do with the Scripps Research Foundation (medical). The Institute has some microbiology, probably marine biology. Its director, Dr. Roger Revelle, is an important figure in the plans for extending the University of California's campus here. Also, he has strong ties with the General Atomic group. As you can see, the Institute [of Oceanography], the new campus, and General Atomic will make a good and functional triangle of research association in a few years.

Meanwhile, Lenora had no job and was quite depressed when, in July 1957, her friend Lorraine Friedman visited her on a vacation. Lorraine, the loyal secretary of Jonas Salk, had much to report to her boss after her first visit to La Jolla. Lorraine noted in her diary her surprise at discovering such a small station when she arrived by train.[3] She obviously expected San Diego to be a big, busy city with a Grand Central–like station. However, she quickly got over her initial disappointment when she discovered how beautiful the city was. Lenora gave her the grand tour of San Diego, and Lorraine was most impressed by Point Loma, the Bali Hai on Shelter Island, and the navy ships in the bay. Lenora then took her to La Jolla and the Scripps Oceanography Institute, which had an expansive aquarium and a pier. Lorraine noted, "This is where Lenora may perhaps get a job with a Dr. ZoBell." Although Lenora was still in want of a job, working at Scripps Oceanography would have been a desperate solution since, as an M.D., she had hoped to work in a medical institution. After her visit Lorraine concluded she loved La Jolla, its beauty, and its weather. Yes, she certainly would be happy to live there. This was fortunate since eventually she did live in La Jolla for forty-two years, until her death in 2005.

Back in Pittsburgh after her vacation, Lorraine quickly had to face reality. At the time, the summer of 1957, Jonas had great hopes of creating his institute on the Pitt campus: the Municipal Hospital had just been renamed Salk Hall, negotiations with Litchfield were progressing, Oppenheimer was supportive, and Dixon was on board. Beautiful La Jolla was better forgotten. However, California was still very much an option, especially in 1959, after the negotiations with Litchfield broke down and Oppenheimer encouraged Salk to leave Pittsburgh and look to the west (see chapter 2).

The obvious choice was the West Coast, and particularly the San Francisco Bay Area. Palo Alto was especially attractive since the new Stanford University Medical Center was being built on the school's Palo Alto campus. The center had been designed so that the Stanford Medical School facilities could be moved from San Francisco to the school's campus, which would transform the medical school into an integral part of the university. The new center included three hospitals and four interconnected medical school buildings. It also introduced an entirely new curriculum and a stellar new faculty. Nobel laureate Arthur Kornberg and most of his department from Washington University had moved from St. Louis, Missouri, to start a new Biochemistry Department at Stanford. Another Nobel laureate, Joshua Lederberg, was heading the new department of genetics. Stanford's move and academic revolution made it one of the most distinguished medical schools in the country overnight. Palo Alto became a highly desirable location for a biological research institute such as the one planned by Salk. Moreover, one of the new hospitals was to be a facility built jointly by the city of Palo Alto and the university. Therefore, the old Palo Alto City Hospital, which was located close to the new center, was to be vacated. On August 1, 1959, various departments of the medical school moved into their new building and the first patients were admitted to the new Palo Alto–Stanford Hospital.

O'Connor was also very supportive of the idea of exploring Palo Alto as a possible location for the new institute. Since April 1959 he had been discreetly investigating what was happening at Stanford. He used as his undercover agent a state representative for the National Foundation–March of Dimes in California, whom he had met on a trip to San Francisco in April.[4] Sworn to secrecy and told to report his findings directly to O'Connor, the agent was to discover all he could about the old Palo Alto hospital that was to be vacated. O'Connor wanted to evaluate the possibility of using the old building to house the new institute, at least temporarily. In a letter dated May 1, 1959, O'Connor's amateur detective reported the results of his investigation.[5] This is an amazing and amusing piece of correspondence. He had made an appointment with the administrator of the hospital under an assumed name: his wife's maiden name. He had collected detailed information about the size and condition of the building: it was large and in good shape. He had inquired about the anticipated date the building would be vacated: August 1959. As to what was to be done with the old structure, that was "still up in the air at this time."[6] The detective assured O'Connor that he typed that letter himself, typing errors and all. He did not make any copies and had kept his mission entirely confidential.

Pursuing the idea of locating the institute in the Bay Area, hopefully in Palo Alto, Salk decided to visit the area himself and make some contacts at Stanford. In preparation for Salk's visit to Stanford, O'Connor wrote a letter, dated July 21, to introduce Salk to Wallace Sterling, the president of Stanford University.[7] O'Connor assured Sterling that the National Foundation had given very serious thought to the establishment of an institute as planned by Salk and that the intention of the National Foundation was very real. Salk visited Stanford sometime in early August 1959, when he had a meeting with President Sterling and probably visited his old friend Clifford Grobstein, who was then a professor of biology at Stanford. Cliff and Jonas had attended high school together in New York before spending four years as classmates at CCNY. Salk also looked up the newly arrived immunologist at Stanford, Melvin Cohn. Cohn was a member of Kornberg's new Biochemistry Department group that had just arrived from St. Louis.[8] Cohn and Salk had met at a few immunology meetings but did not know each other well. Salk, however, wanted to get reacquainted because Cohn was on Szilard's short list of names in the appendix of the Doering-Szilard memo (see chapter 4). They discussed vaccines and probably virus inactivation as well. Cohn's Ph.D. advisor, Alwin Pappenheimer, was an expert on diphtheria toxin and its inactivation to make a vaccine. Salk mentioned to Cohn that he was looking for a location for a new research institute in the Bay Area and had collected information about potential real estate. He knew exactly what he wanted to see, he had a map, he was prepared, and Cohn offered Salk a ride to wherever he wanted to go.

Salk wanted to visit two specific locations in the Bay Area: the Presidio, at the foot of the Golden Gate Bridge, and Treasure Island, a manmade island built under the San Francisco–Oakland Bay Bridge. What these two sites shared in addition to their spectacular settings was the fact that both had been active military bases during World War II. Salk had heard that military bases were being decommissioned and closed, opening up magnificent real estate that had not previously been available for development. He wanted to investigate the potential prospects at those two sites in beautiful San Francisco. As it turned out, those two bases were not scheduled for closure until about 1990, so all Salk and Cohn got from their expedition was a scenic drive and a pleasant talk. More importantly, it gave the two men a chance to get to know each other better.

In hindsight, the reason for Salk's sudden interest in old military installations is clear. In early May 1959, before his visit to Palo Alto, Salk had received a letter and an interesting memo from Szilard.[9] He had not been in touch with Szilard since early 1957, when he had declined an affiliate membership in Szilard's planned institute described in the Doering-Szilard memo (see chapter 4). Obviously, in spite of all his schemes and ideas, Szilard still had not been able to inspire much interest in the realization of his own plans for an institute or been able to raise the necessary funds. He had also heard that Salk was actively negotiating with the University of Pittsburgh to locate his institute on the Pitt campus. In his 1959 letter Szilard inquired whether Salk would consider locating his institute in some place other than Pittsburgh. He added, with his usual bluntness, "Frankly, I see no possibility of getting many first-class people to move to Pittsburgh."[10] He went on to explain that his friend Roger Revelle, the director of the Scripps Institute of Oceanography, had been put in charge of developing a University of California campus in La Jolla.[11] Revelle had asked Szilard to inquire whether Salk would consider locating his institute in La Jolla. Szilard added, "I know La Jolla very well and think it is an excellent place." Little did Szilard know that Salk had already decided to locate his institute in California and was considering Palo Alto. As for the University of California campus planned by Revelle, that was old news to Salk. He had heard all about it from Lenora and Lorraine.

The 1959 memo enclosed with the letter was more interesting to Salk. It reported Szilard's recent conversations with Jim Watson and Roger Revelle about the possibility of convincing Salk to set up his institute in La Jolla. It included the names of desirable people that might join the institute and a statement that aroused Salk's curiosity: "Revelle thought that it would be comparatively easy to obtain at La Jolla a tract of land, as a gift, upon which such an institute could be built." Salk, however, did not take the availability of this free land for granted. He was too familiar with Szilard's wild imagination and unrealistic expectations. Salk and Revelle, however, spoke on the phone in June. Revelle must have confirmed that free land had indeed become available as a result of the closure of military camps after World War II. The land, which had been leased to the military during the war, belonged to the city of San Diego. General Atomic had been the first beneficiary. The city had also offered free land to develop the new University of California campus.

If land from decommissioned military facilities was becoming available in San Diego, why wouldn't the same be true in San Francisco? In late June, after his telephone conversation with Revelle, Salk dictated a note to Lorraine: "We talked about the possibility of my going out there; I would write to him if I were so interested. I believe I will write in the affirmative and use this as an excuse for moving ahead on the Palo Alto situation. . . . I also want to speak to Lenora Brown. Would you think of time for such a trip for me?"[12] Acting on the information he got from Revelle, Salk must have decided to investigate whether such a gift of land might be an option in the Bay Area. He easily identified the two large military installations in the San Francisco Bay Area that had played a major strategic role during World War II: the Presidio and Treasure Island. He then organized his trip to Palo Alto, obtaining the letter of introduction from O'Connor to Sterling and preparing for Sterling a document entitled "Purpose and Implementation of a Proposed Institute."[13]

Immediately after his return from Palo Alto, Salk sent a letter—dated August 6—to the renowned broadcast journalist Edward R. Murrow, his friend and confidant.[14] Salk had met Murrow in 1955 when he was interviewed on February 22 for the popular TV show *See It Now* (see chapter 1), a show that propelled Salk to the forefront of the polio crusade. A few weeks later, on April 12, Murrow hosted the famous live broadcast from Ann Arbor (see chapter 1), which featured Salk and Dr. Thomas Francis after the announcement of the success of the vaccine. It was largely Murrow's talent that made Salk a hero in the eyes of the public. Salk and Murrow had since developed a lasting friendship based on great mutual respect and trust. Salk considered Murrow his personal advisor and consulted him about the creation of his institute. Salk's letter written after his return from Palo Alto is revealing of their relationship: "Nothing definitive will be done nor steps taken without getting your reaction."

He enclosed to Murrow a copy of O'Connor's letter to Sterling, adding, "We are awaiting his reaction." Salk went on to review the names of potential members of the board of trustees that he and O'Connor had agreed upon, including Murrow himself. Salk then relayed his report of recent events to Murrow:

> I don't know whether or not I have mentioned to you that I have just had a firm invitation to set up an institute, such as this, at La Jolla, adjacent to the new University of California campus, intended at the moment for a Graduate School but, perhaps, eventually for other activities as well. Harold Urey is there and a number of other good people are being attracted. This clearly is not my first choice. I feel Stanford is the place and I am hoping the feeling I had from Sterling, of interest on his part, will be matched by people at the working level. . . . I hope we may hear from Sterling before you leave the country and in sufficient time to meet with you, even if but briefly. . . . I plan to be away from the laboratory for the next three weeks but will interrupt my vacation for anything that concerns the matter of the institute.

Salk did interrupt his vacation for his first trip to La Jolla, where he spent from August 17 to 21, 1959.[15] He was wined and dined by Revelle and his gracious wife, Ellen Clark Revelle, who was quite used to entertaining dignitaries at her spectacular home bordering the white sandy beach of La Jolla.[16] Her warm hospitality had attracted many UCSD faculty members. A remarkable woman, she was a Scripps family heiress, and her great-aunt was Ellen Browning Scripps, the benefactor of La Jolla whose generosity included founding the Scripps Institute for Oceanography and the Scripps Metabolic Clinic. The Revelles' famous hospitality worked on Jonas Salk. Ellen drove Jonas around herself so he could tour various sites, and Roger proudly showed off the location of the future University of California campus, *his* campus. Salk was made to feel welcome and among friends.

Immediately after Salk's departure, Revelle wrote to University of California president Clark Kerr to report on the visit.[17] Salk's impression had been right, and Revelle's letter to Kerr was enthusiastic. He described Salk as "modest, serious, idealistic and experienced," and he confirmed Salk's impression that, "on our side, we like him a great deal both as a man and as a scientist." Revelle thought that the La Jolla campus would greatly benefit from the proximity of the institute proposed by Salk. It would help in the development of science and engineering and of the future medical school. Revelle also mentioned to Kerr that Salk was considering Stanford and had visited Sterling. He urged Salk to talk to Kerr as soon as possible and asked Salk to arrange for him to meet O'Connor.

Upon his return to Pittsburgh Salk was clearly shaken; he had much to think about. He had been touched and impressed not only by La Jolla's beauty and the friendly welcome he had received, but even more by the size, scope, and vision of what was happening there and the possibilities offered for the future. The available land not only had a superb location, but it was also large and mostly empty. It was located on a high plateau that, at one point, reached the coast ending in a bluff plunging into the ocean. The site was called Torrey Pines Mesa because it was adjacent to a state park, a natural reserve covered with the nation's rarest pine tree, the spectacular wind-sculpted Torrey pine. The trees were so striking that the first name Salk chose for his institute in La Jolla was The Institute for Biology at Torrey Pines.

Torrey Pines Mesa was part of San Diego's "pueblo lands." In 1834 Mexico had recognized San Diego as a pueblo and granted it the right to municipal government and, with it, the right to pueblo lands. Amazingly, when California became a U.S. state, it was because it had been a Mexican pueblo that San Diego inherited the legal rights to land and water. The ownership of those valuable lands is the "Hispanic dowry" of the city of San Diego.[18] By 1874, after much legal haggling, San Diego received the patent to about forty-eight thousand acres of pueblo lands. That land had been surveyed and divided into well-defined and numbered 180-acre lots on Torrey Pines Mesa. Thanks to wise restrictions on the sale of pueblo land, some land owned by the city of San Diego was still available in the twentieth century. It was this land that lured suitors to develop the economy of San Diego, as large parcels were either gifted or sold at favorable rates. An early suitor was a railroad company, and later those lands were sold to establish military, educational, business, industrial, or residential developments. In the case of temporary installations, such as military camps in wartime, city lands were leased to the federal government for a nominal sum. The transfer of ownership of pueblo lands requires that voters authorize the city council to convey the land if the gift would be of benefit to the city and its people. A simple majority vote for approval is sufficient, except in the case of parklands, when a two-thirds majority is required for passage.

About 1,300 acres of Torrey Pines Mesa were leased to the United States military in 1940. In January 1941 the site opened as a Coast Artillery Corps Replacement Training Center for new inductees. The purpose of this training center, which went by the name Camp Callan, was to prepare soldiers for a Japanese attack of the California coast by sea or air. After Pearl Harbor in December 1941 this had become a real threat, especially since San Diego had expanded its aircraft plants and had both a powerful naval air station and a Marine Corps base. War goods, especially planes built in San Diego, were being rushed to England to protect Europe, but the United States was unprepared for an invasion along its southern Pacific coast. Moreover, San Diego was the repair and operating base for the aircraft carriers that had escaped the disaster of Pearl Harbor. Suddenly, the protection of the coast—and Camp Callan—seemed long overdue. By March 1942 Camp Callan had become a hive of activity. About fifteen thousand trainees went through a thirteen-week training period here, where they learned to use all kinds of firearms, from rifles and pistols to big 155 mm guns and antiaircraft weapons. Today it is hard to imagine those big guns firing heavy shells into the ocean from the Salk Institute parking lot.[19] Torrey Pines Mesa became a small town with about three hundred barracks, a thousand-bed hospital, three theaters, five chapels, support and storage buildings, and a landfill. Camp Callan eventually got its own weekly newspaper and military band.

With the end of World War II the camp became pointless: it was decommissioned, declared surplus in November 1945, and disman-tled.[20] With the departure of the military, La Jolla went back to being a small, sleepy town, primarily a family summer resort and a comfortable community for retirees. Nothing remained of Camp Callan except some ruins and the foundations of the barracks. Torrey Pines State Park was undamaged by the military and remained open to the public. In the mid-1950s about one hundred acres of what used to be Camp Callan, south of the park, were set aside for the construction of a spectacular public golf course. The city-owned glider airport, the Torrey Pines Gliderport, also south of the park, had become part of Camp Callan during the war, but it reopened to the public in 1946 and full-scale sailplanes returned. The cliffs along Torrey Pines Mesa had been used for soaring by aviators since at least the 1930s. The most famous of them, Charles Lindbergh, flew along the 350-foot cliffs from Mount Soledad in La Jolla to Del Mar in 1930.

When Salk first visited La Jolla in August 1959, the only new building on the mesa belonged to General Atomic, which had been founded in July 1955 as a division of General Dynamics, a major defense contractor. General Atomic had been the pet project of the chairman of General Dynamics, John Jay Hopkins, who wanted to explore postwar uses of atomic energy. Hopkins recruited Frederic de Hoffmann as the first general manager of General Atomic.[21] Freddie, as he liked to be called by his colleagues, was then a vice-president at the Convair Division of General Dynamics in downtown San Diego. He was a nuclear physicist who, in 1943, as a twenty year old, had been picked out of Harvard to work in Los Alamos on the Manhattan Project. Together the two men started dreaming about "Atoms for Peace." They had the idea, the expertise, and the money; all they needed was a location.

The timing of the founding of General Atomic was perfect. In 1955, after sitting on the San Diego City Council for twelve years, Charles C. Dail was elected mayor. San Diego was suffering from a postwar recession and needed to expand into new areas in addition to tourism and the defense industry. Mayor Dail had a vision for San Diego and was full of energy and ambition, in spite of a physical handicap: Mayor Dail was a polio survivor. He was eager to develop and diversify the San Diego economy by attracting new businesses and industries, but he did not want his city to become another polluted Los Angeles. He wanted to attract clean businesses and organizations that would create well-paid jobs, and he had the land to do so—the pueblo lands of La Jolla. General Atomic seemed to qualify, and, in the fall of 1956, the city voters approved the transfer of 320 acres of Torrey Pines Mesa, just east of the state park, to General Atomic. The new company spent three years in a temporary location, a school near the San Diego airport loaned by the city. The new General Atomic building was then dedicated as the John Jay Hopkins Laboratory for Pure and Applied Science, in recognition of the contribution of Hopkins, who had died in 1957. The dedication took place on June 25, 1959, only a few weeks before Salk's first visit. People were still talking about the impressive ceremony presided over by Niels Bohr, who came all the way from Copenhagen to give the keynote speech. It was

Bohr who had then thrown the switch to present General Atomic's first product: a safe nuclear reactor named TRIGA.[22]

TRIGA was the creation of a group of about ten people, many of whom had been involved in the Manhattan Project. The safe reactor team included Edward Teller, Freeman Dyson, and, of course, Massoud Simnad.[23] Edward Creutz was the first director of research at General Atomic. He had become head of the Physics Department at the Carnegie Institute of Technology in Pittsburgh after the war. That is where he discovered Massoud. It turned out that Creutz was also a close friend of Leo Szilard. They had both worked on the Manhattan Project in Chicago, but Creutz had eventually moved to Los Alamos. It was Creutz who, in 1945, had introduced Szilard's petition not to drop bombs on Japan into the Los Alamos compound since Szilard was banned from entering the facility (see chapter 3). Years later Creutz invited Szilard to be a consultant for General Atomic, and Szilard spent two months in La Jolla in early 1959. That is how Szilard discovered La Jolla. As far as the new General Atomic building was concerned, it was fabulous and futuristic, circular and reminiscent of a huge flying saucer or a cyclotron.[24]

While General Atomic already existed in La Jolla, the future UC campus was only in the planning stages, but those plans were impressive and in motion. The first faculty members had arrived in 1958 and included the chemist Harold Urey, from the University of Chicago, another good friend of Szilard. Urey had trained with Niels Bohr, had won the Nobel Prize in Chemistry in 1934, and was a Manhattan Project veteran. Expected in 1959 was the theoretical physicist Keith Brueckner, recruited from the University of Pennsylvania, who had been attracted to La Jolla by the General Atomic group working on the new nuclear reactor. Due to arrive from Yale in 1960 was David Bonner, a geneticist trained at Caltech by George Beadle and Edward Tatum. In 1958 Beadle and Tatum had shared a Nobel Prize that recognized work for which Bonner claimed considerable credit. The graduate School of Science and Engineering had been authorized in 1956 by the University of California regents and was to be temporarily housed in the Scripps Institute of Oceanography (SIO) building. Only in May 1959, after years of negotiations and maneuvering, had the regents approved La Jolla as a site for the development of a new UC campus.[25]

All this was largely the work and dream of Roger Revelle. In 1951 he had been appointed director of the SIO, which was a part of the University of California system and where graduate students could work toward a Ph.D. that would be granted by UCLA.[26] Revelle was, therefore, the chief local administrative officer of the University of California. In 1956 he had been put in charge of exploring the possibility of creating a branch of UC in the San Diego area. This could be conceived as an extension of the SIO facility into a graduate program in science and technology or as a new campus. In the Cold War climate of the time, San Diego—a city that was favorably inclined to the military, committed to defense, and had the support and encouragement of the new General Atomic—welcomed a graduate school to educate a new generation of scientists and engineers. After Sputnik, according to common advice repeated by Revelle, "The nation's youngsters must learn either science or Russian!"[27]

As for a new campus, that was justified by the expected increase in the population of California after the war. The UC regents retained Charles Luckman, of the architectural firm that built General Atomic, to advise on the site for the campus. He recommended the land north of SIO and adjacent to General Atomic with two conditions: that at least a thousand contiguous acres of land could be obtained as a gift and that a city master plan would assure the development of a community supportive of a large campus. Fulfilling those conditions was incredibly complicated. It required not only the acquisition of land through gifts, purchase, and swapping, but also moving highways, building new access roads, and planning for commercial and residential developments. Moreover, Revelle had very clear ideas about the kind of campus he wanted, and his plan involved integrating the SIO into the new campus. The voters had to approve the transfer of all or part of a whole series of pueblo lots in order to link the SIO and the future campus. All this had to happen in spite of the hostility of some powerful regents who favored UCLA or UC Berkeley because they were concerned that the new San Diego campus would compete with older UC campuses for faculty, students, and resources.

The timing of Salk's first visit to La Jolla, August 1959, was remarkably lucky. At that time the future UC campus was essentially approved, but the details were still being discussed, and the precise boundaries and academic future of the new campus were still unclear. This implied flexibility, as many options were still open. Salk had visited La Jolla only to eliminate it as a possibility so he could concentrate his efforts on his favorite location, Palo Alto. Upon his return to Pittsburgh, however, he was disturbed. As was typical of Salk, he patiently compared Palo Alto to La Jolla.

PALO ALTO	LA JOLLA
Full intellectual Community	to be developed
Full medical complex	to be developed
Good basic biology	special fields
Interesting Behavioral Sciences Center	o
Proximity to San Francisco	San Diego
" " Berkeley	Los Angeles & Pasadena
Very good climate	excellent climate
Some adaptation to local attitudes	Pioneering (two different kinds of self confidence)
Uncertain desire by medical community	Strong desire by university and popular committees
Possibly better general appeal to scientist[s] and intellectual[s]	Probably not appealing to all
More certain survival in the intellectual environment	Intellectual strength influenced by university potential
More certain prompt effects (through ERM [Edward R. Murrow])	Possibility for rapid start less likely
Land and money not too much of a problem	Land and money for building easier to get evidently[28]

The comparison was realistic, systematic, and thorough. Other handwritten notes by Salk dated August 26 and 27 further reveal his thoughts and concerns. He worried about the intellectual isolation of La Jolla: the limited number of scientific colleagues and the lack of intellectual stimulation in general. He clearly wanted to keep both La Jolla and Palo Alto open as options for his institute and awaited the next move from Palo Alto. He mentioned also a broader interest for the institute itself, jotting notes about a planned Institute for Biological and Humanistic Studies. He was uncertain of his own judgment, however, and decided to contact a number of people to get their reaction. Meanwhile, on September 14 in New York he met with UC president Kerr, who essentially confirmed the plans for the La Jolla campus.

In September Salk drafted a lengthy letter to Massoud and Lenora, not only thanking them for their kindness during his visit but also sharing his impressions of La Jolla and his sudden wavering between Palo Alto and La Jolla. He admitted that he was very disturbed by the La Jolla experience, saying, "It is only fair for me to tell you that the order of preference at this moment has shifted." The draft of that letter is much more interesting than the letter he actually sent off, which was merely a short thank-you note.[29] Salk scrapped most of the first draft as if he was afraid of what he had confided. He had dictated the initial letter as if he were talking to himself while traveling to his cottage in Maryland, and he may have been somewhat shocked by what he had revealed when he later read the transcript.

In October Salk made numerous plans to talk or meet with the men on the top of his list of those he might want as original members of such an institute. These men included Seymour Benzer, Melvin Cohn, Renato Dulbecco, Herman Kalckar, Matthew Meselson, Theodore Puck, Leo Szilard, and Jim Watson.[30] He also consulted with Hermann Muller at Bloomington and with George Beadle at Caltech.[31] As he later wrote to his confidant, Ed Murrow, "I was astonished that all of them saw the greater possibility of La Jolla over Palo Alto, even though I withheld my own views until they expressed theirs." Salk also appreciated "the opportunity for pioneering, in a way, and doing something quite different without being surrounded too tightly by the past and by academic prejudices in an already established university community." Salk also commented, "The site is unbelievably beautiful—at the moment it is intended to be a park area and, therefore, will require a two-thirds vote of the city at an election in June to make tentative agreements final. I understand this is a formality." In this, Jonas was a little naïve.

Salk's second visit to La Jolla took place November 14 to 16, 1959. This visit was more official than the first one, and Basil and Mrs. O'Connor accompanied him. Reporters photographed them as Revelle showed them the site of the future campus.[32]

Salk's third visit, from January 2 to 11, 1960, began auspiciously as Salk clarified that "he wanted the institute to be independent of the university but located near it and cooperating with it in all possible ways."[33] This idea was well received by Revelle, who was relieved at not having to fit the institute within the university, as he had anticipated possible problems with that plan. However, it was also during this visit that the relationship between Revelle and Salk started to crumble as the so-called Salk Affair began.

On that visit Salk met San Diego's city fathers, including Mayor Dail, the polio survivor who must have been thrilled to meet his hero. The city officials showed him the potential sites for the institute, including the parkland to the west of Pueblo Lot 1324. That lot included Torrey Pines Park and the golf course, but a portion of the lot, between the golf

FIGURE 1. Roger Revelle (center) showing Jonas Salk (left) and Basil O'Connor (right) the future location of UCSD. November 1959. *La Jolla Light.*

course and La Jolla Farms, was apparently still available—and Jonas fell in love with it. What was special about it was that it was the only city-owned lot located directly at the edge of the mesa. Some of the land, such as the strip of land that included the delicate bluff itself, was protected parkland, but the city officials informed Salk that some flatland in lot 1324 might be available. Unfortunately, Revelle believed that the area had been reserved for the university in a previous master plan, although no precise boundaries had been specified. Revelle considered that spectacular site to be the heart of his campus. He tried to offer Salk an alternative: William Black, an oilman and real estate developer, had purchased the pueblo lot above SIO in 1949 and developed it as La Jolla Farms. He still owned vacant ocean-view land at the south end of La Jolla Farms. Revelle suggested over dinner with Black that he might sell or give up that property for Salk's institute, but that idea went nowhere.

When Salk returned to La Jolla the first week of February, he came with moral support. O'Connor came along, as did Louis Kahn, a Philadelphia architect he had recently met (see chapter 8). Salk informed

Revelle that he was negotiating with the city for all of the available land in Pueblo Lot 1324, parkland included. Kahn was to give him advice for a tentative layout of the buildings to submit to the city council in March. Salk then spent a week in La Jolla in March to present his request for land at hearings of the city council.[34] Lorraine, who accompanied him on that visit, recalled that, while in the car with Salk, she remarked, "You are getting land and planning a building but you don't have a cent in your pocket," and Salk replied, "Right. But they don't know that."[35] Jonas was not bluffing: he sincerely trusted O'Connor's promises and his power to deliver. O'Connor also addressed the city council, officially stating the financial commitment of the National Foundation to the new institute: one million dollars a year for ten years toward an endowment, and one million annually to support the operations.[36] Nothing was said about the source of the funds to pay for the building.

On March 9 the local newspapers reported that the city intended to give Lot 1324 to Salk, and Revelle delivered a furious letter of protest to the mayor. The conflict became very public and unpleasant. Salk was so popular that Revelle and the university were largely blamed: "Saying no to Jonas Salk was like saying no to Apple Pie and Motherhood."[37] It was believed that even some UC regents supported Salk. However, the voters were not ready to give away more parkland.[38]

Eventually a solution was found that avoided distributing parkland entirely. Proposition E asked voters to release forty-six acres of ordinary land from Lot 1324 and to authorize transfer to either Salk's group or the university. An agreement would then be reached—behind closed doors—to split that parcel between Salk and the university. However, some campaigning would be necessary before the proposition would pass, even though its approval was assured before the unpleasantness of the Salk Affair. Mayor Dail organized a Citizens Committee to back Proposition E on the June 7 municipal ballot. Interestingly, Frederic de Hoffmann was the honorary chairman of the Committee for Advanced Education and Medical Research. The immodest slogan spread in big letters at the bottom of that committee's stationery was "Let's Make San Diego the Scientific Capital of the World."[39]

Proposition E eventually passed with a comfortable margin and the parcel was split into two halves of about twenty-seven acres each, both at the edge of the bluff. A road was built between the two halves, allowing public access to the parklands and glider port.[40]

Whatever happened to the Palo Alto idea? In April, before the municipal ballot of June 7, Salk received a confidential handwritten note from his friend Ted Puck.[41] At a meeting in Chicago Puck had met Henry Kaplan, a highly respected radiation oncologist, a pioneer in the treatment of Hodgkin's disease, and chairman of the Department of Radiology at Stanford. He had been a strong supporter of the move of the Stanford Medical School to Palo Alto and its development.[42] He had worked closely with Sterling, Stanford's president, and was well informed about the university's politics. Puck wanted Salk to know that, according to Kaplan, Stanford was still an option. Kaplan mentioned three reasons for the difficulties that arose during the negotiations. Some resented Salk's attitude that because the university was contributing no money, it should have no relations with this institute. Others doubted that Salk could attract the people he mentioned, or were unconvinced of the financial soundness of such an institute. Kaplan thought that a positive conclusion could still result if Salk resumed the negotiations and addressed those concerns. However, that never happened, and the relationship with Stanford petered out. In any case, Stanford already had attracted many big names: they did not need Jonas Salk to promote their campus. The La Jolla campus would benefit by attracting Salk. That was where Jonas was welcome and wanted, and where his institute would end up.

On December 27, 1960, Salk wrote a very brief letter to Cohn:[43]

Dear Mel:
    Lest you think that the New Year has nothing new in store—
    I send you the enclosed.
    I look forward to seeing you in La Jolla.
    As ever,
    Jonas

Enclosed were the articles of incorporation and first bylaws of "The Institute for Biology at Torrey Pines (a California nonprofit corporation)."

CHAPTER 6

The Pasteur Connection

Botticelli is a wine, not a cheese, you fool!

—*Punch*

The Pasteur connection to the Salk Institute was initiated in the summer of 1946 at the Cold Spring Harbor symposium entitled "Heredity and Variation in Microorganisms," the first CSH symposium after World War II (see chapter 4 and frontispiece). This was a seminal meeting that launched the field of bacterial genetics. Its remarkable audience included several Frenchmen, notably Jacques Monod and André Lwoff from the Institut Pasteur in Paris. It was at that meeting that Monod met Alwin Pappenheimer, who was then a professor of microbiology at New York University. After the meeting Monod and Lwoff spent the weekend at Pappenheimer's summer house in Scotland, Connecticut, where they became friends. By the end of that weekend, Monod, Lwoff, and Pappenheimer had turned into Jacques, André, and Pap—the nickname used by all of Pappenheimer's friends (see frontispiece).

Alwin M. Pappenheimer Jr., the oldest of three children, was born in Cedarhurst, New York, in 1908.[1] His father, a well-known pathologist on the faculty at Columbia University, had a great influence on his children, all of whom became scientists and professors at Harvard. They were brought up in a sophisticated academic environment. Theirs was also a very musical family, with Pap playing the clarinet and the viola. In 1919 the family moved to Hartsdale, where Pap attended the Lincoln School of Teachers College. At seventeen he entered Harvard, where he was the first student to enroll in the newly created biochemical sciences tutorial. He wanted to prepare for a career in biological research but realized early on that a solid background in chemistry and physics was essential. Rather than attending medical school, he did graduate work in organic chemistry at Harvard and obtained a Ph.D. with James B. Conant in 1932. He did postdoctoral work at Harvard in Hans Zinsser's Department of Bacteriology and Immunology. While Zinsser was developing a vaccine against typhus, John F. Enders, then a graduate student in Zinsser's laboratory, was working on pneumococcal polysaccharides. Pap joined him on that project, and it is this early work that brought him to the attention of Oswald Avery at the Rockefeller Institute. This, quite by chance, would later greatly affect Pap's career.

After two years at the National Institute of Medical Research in London, he returned to Cambridge, Massachusetts, in 1935—in the middle of the Great Depression. He still had no job, but he had chosen the problem that interested him. Throughout his career he would pursue this interest: isolating and understanding the mode of action of a bacterial toxin. Still on a fellowship and working in a borrowed space at the Antitoxin and Vaccine Laboratory in Jamaica Plain, a neighborhood of Boston, Pappenheimer isolated and characterized diphtheria toxin.[2] This was the first bacterial toxin to be fully purified, and the work brought him recognition and employment. In 1939 he accepted a position as assistant professor of bacteriology at the University of Pennsylvania.

In 1941 Pap accepted the offer of an assistant professorship at the NYU School of Medicine. The attraction at NYU was Colin MacLeod, who had just become the new chairman of the Department of Bacteriology and been given the opportunity to recruit new faculty for the new Department of Microbiology. It was Oswald Avery who, remembering Pappenheimer from his early work on pneumococcus in Zinsser's laboratory, had recommended him to MacLeod. Avery and MacLeod, who had been working together at the Rockefeller Institute, had just made the very interesting observation that an extract of the virulent S strain of pneumococcus could transform an avirulent R strain into the dangerous, virulent S strain.[3] Pappenheimer and MacLeod agreed on changes that would be necessary in order to build an exciting department. Their plans, however, had to wait because the United States entered World War II after the attack on Pearl Harbor in December of that year. Pap became a captain in the Medical Corps engaged in the South Pacific. Through a series of extraordinary circumstances, his military service turned out to play a key role in the creation of the Salk Institute. He did not return to New York until late 1945, only a few months before the June 1946 CSH symposium where he met Jacques Monod.

Jacques Monod, the youngest of the three sons of Lucien Monod and Charlotte Todd McGregor, was born in Paris in 1910.[4] The Monods belonged to an old family of French Protestant Huguenots that could be traced back to the fifteenth century. Victims of religious persecution, many Huguenots left France in the sixteenth and seventeenth centuries. Jacques Monod's ancestors had immigrated to Geneva, but one of them, Jean Monod, eventually returned to France in 1808, where he recovered his French nationality while remaining a citizen of Geneva, where he was born. Not surprisingly, the people who objected to the established religion did not represent the average population. They were educated and successful professionals, often pastors, professors, scientists, civil servants, or bankers. Jacques Monod's father was unusual in that he was a painter as well as a scholar. His mother, an American born in Milwaukee, was the granddaughter of a pastor who had emigrated from Scotland. She came to Europe as a tourist and settled in Paris. When she was looking for a painter to make her portrait, someone recommended Lucien Monod. Lucien must have been a charmer in spite of a major handicap: he could walk only with the help of crutches, a serious sequela of having had polio as a child. Moreover, Lucien did not speak a word of English, which he always refused to learn. It is Charlotte who learned to speak French with an amazing American accent. As a result, the Monod children were perfectly bilingual.

Sadly, their firstborn, Robert, suffered a disastrous accident at birth that left him mentally impaired. Their second son, Philippe, born in 1900, was vigorous, smart, and handsome.[5] Jacques was born ten years later. By 1914, the First World War had made life as an artist in Paris difficult, and Lucien Monod returned to the roots of his family—Switzerland. Geneva had become the neutral refuge for the rich, and Lucien obtained a few commissions to paint portraits. Although they were safe living in exile, making a living was hard. Then, quite suddenly, Jacques became ill. He woke up one morning with a high fever and his left leg paralyzed. His father had no trouble recognizing the disease that had crippled him as a child: polio. Luckily, Jacques recovered quite well. The disease left him with a somewhat shorter leg and a slight limp that could largely be corrected with a special shoe. Like most polio survivors, Jacques was not prevented from practicing the challenging sports he loved, mountain climbing and sailing. One must also wonder whether his family history with polio may have been a source of his respect and affection for Jonas Salk.

After World War I Lucien and his family settled in Cannes, where he rented a house in 1918. He wanted to raise his sons in a town that was smaller and less hectic than Paris, in the attractive light and climate of the south of France. Since Lucien had come into an inheritance from the Monod side of the family in 1917, life had become easier. By 1919 he bought a large villa, Le Clos Saint-Jacques, above Cannes, overlooking the Mediterranean Sea. This is where Jacques was raised mostly under the influence of his father, his first mentor, who was not only a gifted and sensitive artist but also an erudite. Upon reading Darwin, his father got Jacques interested in biology and philosophy at a young age. Music was also a major activity in the Monod family. Jacques learned the piano and the cello, instruments for which he obviously had a talent. He attended a junior college in Cannes and obtained his *baccalauréat* in philosophy, not science.

Interested in the mysteries of nature, he left Cannes in 1928 to study the natural sciences—general biology, chemistry, zoology, and geology—at the Sorbonne. In Paris, where he shared the apartment of his brother Philippe, he felt like a transplanted meridional. As a student in zoology, he spent his summers at the marine station of the University of Paris in Roscoff, on the north shore of Brittany. It was in Roscoff that he eventually met the four scientists who, although they were not his professors, truly initiated him to biology: Georges Teissier revealed to him the power of a quantitative description of biological phenomena; André Lwoff encouraged him to discover microbiology; Boris Ephrussi introduced him to the field of genetics; and Louis Rapkine gave him the idea that life has to be described at the chemical and molecular levels.

Monod graduated with a *licence ès science* from the University of Paris in 1931. He clearly wanted to do research and obtained fellowships to receive training in France, first in Strasbourg and then in a laboratory at the University in Paris. In 1934 a position opened up as an assistant zoologist at the Sorbonne, an opportunity so rare that he could not afford to pass it up. However, that summer he left the laboratory as the naturalist for a polar expedition with the famous French explorer Jean-Baptiste Charcot. Sailing on the *Pourquoi-Pas?* opened a new world to Monod, who could not wait to join another such expedition in 1936. However, by then he had been offered an even more tempting adventure: Ephrussi convinced Monod to join him at Caltech in Pasadena, California.[6] Thanks to Rockefeller Foundation fellowships, Ephrussi and Monod were able to join the group of Thomas Hunt Morgan for a year.

At Caltech Monod came to realize how backward the scientific education he had received at the Sorbonne was. He recognized the importance of genetics as a discipline and discovered an entirely new atmosphere in the laboratories—so different from the stuffy Sorbonne—with informal discussions, friendly contacts, and cooperation between colleagues. Initially he tried his hand at some experiments with flies, but, after returning to Paris in 1937, he started the quantitative studies of bacterial growth for which he was awarded his Ph.D. in 1941. Meanwhile, Nazi Germany had invaded France in May 1940 and Paris was occupied. Monod joined the underground and, as it became too dangerous to go back to the Sorbonne, he started doing some experiments at the Pasteur Institute.

After Paris was liberated, Monod joined the staff of General de Lattre de Tassigny to integrate the various resistance groups into the First French Army.[7] During this time he met American officers and had access to American scientific publications, which is how he discovered the Luria-Delbrück and Avery-MacLeod-McCarty papers (see chapter 4). He again joined the Sorbonne Zoology Lab, where nobody was interested in his work. Lwoff then offered him a position in his Service de Physiologie Microbienne and encouraged him to attend the 1946 Cold Spring Harbor Symposium, where he met Pappenheimer.

Monod and Pap certainly had much in common: they were about the same age, they both came from a distinguished family of gifted scholars, and both were talented musicians. They even had the same approach to science. Although they had broad scientific interests, they both picked a biological problem early in their career and pursued it for decades until it was solved at the molecular level. Both were inspiring and devoted teachers who left as part of their legacy a large school of accomplished and grateful students. Monod and Pap even had a student in common: Melvin Cohn.

Melvin (Mel) Cohn was born in Manhattan in 1922. His grandparents had immigrated to the United States to escape pogroms in the Russian Empire in the late nineteenth century. Both of his parents graduated from Brooklyn Law School in 1918. His mother was one of the first female lawyers in New York, becoming a lawyer before women had the right to vote in the United States. She was also the first of three generations of women lawyers in the Cohn family, as Mel's sister and her daughter also became lawyers. Mel's father was most unusual in that he became blind at a very young age. The cause of his blindness is uncertain, but it was the result of disease, not an accident. Unable to

FIGURE 2. Jacques Monod in the First French Army, 1944.
Courtesy Pasteur Institute Archives.

care for a blind child, his parents had him raised first on a farm and then in an institution for the blind, where he learned to read Braille and play the organ. He also turned out to be a very talented pianist. Because of his blindness he developed an extraordinary memory, remembering anything he had heard once, whether it was related to law or music. In the 1920s, Mel's parents held a private law office and were quite well off. However, business slowed during the Great Depression. Fortunately, Mel's mother got a job with the federal government in the housing department handling foreclosures, which were rampant during the depression.

Mel had to earn fellowships and pocket money throughout his education. While he attended Boy's High School in Brooklyn, he held odd jobs at night. During the summer, he sold candy in a burlesque theater and danced the lindy hop in a zoot suit at a hotel. That particular job he remembers well because the advertising agency that hired him to show off clothes to retailers paid him a royal fee: ten dollars, "enough to pay the subway fares for months." He then attended the City College of

New York and graduated with a bachelor's degree in physics in 1940. During his college years he held a night job at the CCNY library (reshelving books) and a summer job with Superior Printing (mixing inks to match colors to print posters). He then entered Columbia Teachers College, where the tuition was very low, perhaps hoping that teaching credentials might get him a job somewhere. More importantly, being a student at the Teachers College greatly reduced the tuition for science classes at Columbia. Although his bachelor's degree was in physics, he wanted to learn more chemistry, and in 1941 he obtained a master's degree in colloid chemistry. At the time, the study of colloids included the study of proteins, and one protein he studied in Columbia's chemistry laboratory was keratin. This was a lucky break that landed him a job as a chemist developing formulas for hair coloring in the research division of Clairol in Stamford, Connecticut.[8] While he was working for Clairol and taking night classes at Columbia, the United States entered World War II.

In early 1943 Mel was drafted and reported to Fort Dix, New Jersey, where he received basic military training.[9] His laboratory experience in chemistry gave him a choice of two types of laboratory assignments: either in an explosives and poisonous gases unit or a medical unit. The choice was not hard for Cohn, and he entered biological research as a member of the Army's 26th Medical Company. His bachelor's and master's degrees sent him to Officer Candidate School at Fort Benning, Georgia. This is where his troubles started. He simply did not understand how the army worked and how he was supposed to behave. For his entire life, at home and in school, he had been raised to respect reason and had been encouraged to ask questions. Such habits are hard to break. In the army, so many things appeared irrational, and he had not learned to quietly follow orders that did not make sense to him. He quickly antagonized his superiors, who declared him unfit to be an officer, and he flunked officer school for alleged "dumb insolence." He was sent back to Camp Ellis, Illinois, to receive training in standard clinical laboratory procedures before being shipped, as a private, to the South Pacific Theater of Operations.

On July 23, 1944, he embarked on a troopship from New Orleans, and, after traveling through the Panama Canal, he ended up in southern Australia on August 26. The trip, which involved rough sailing in equatorial waters during the stifling summer heat, was hell. Most of the men were seasick, traveling belowdecks in hammocks hanging over filthy seawater sloshing in the stinking hold. It was announced that an officer was to teach a class in elementary chemistry to selected medical lab technicians. Mel, in spite of (or perhaps because of) his scientific training, was selected to take the class, which was held in the hold along with Mel's essentially uneducated classmates. Mel would always remember that the teaching officer started the course by emphasizing the importance of acid/base balance and writing on the blackboard the Henderson-Hasselbalch equation defining pH. The men were baffled, with no idea of what this meant. Mel was delighted—finally, something that made sense! After class he introduced himself to the teacher and found out his name was Captain Pappenheimer. Mel had no problem interacting with an army officer who was a professor at NYU, and they got along right away. Pap made the rest of the trip not only bearable but even interesting. Mel learned much on that trip about bacteria, enzymology, and diphtheria toxin.

After disembarking on the south coast of Australia in August 1944, Mel met a young officer who had graduated with bachelor's and master's degrees from the University of North Carolina in 1940: Lieutenant Lane Barksdale. He was in charge of a small unit within the 26th Medical Company that specialized in the diagnosis and control of tropical diseases such as malaria, leprosy, filariasis, schistosomiasis, and amoebic dysentery, among others. Lane got Mel to join his group, and this was to be an extraordinary learning experience as they spent the rest of the war together. Pap, on the other hand, was not part of the field unit and left them to join a military hospital.

Before learning about the local tropical diseases, however, Lane and Mel had to learn to fight. Usually medical corps personnel do not carry weapons, since they keep to the rear and are protected by the Geneva Conventions. However, the 26th Medical Company was a mobile laboratory that worked on the battlefield, scouting the area for deadly pathogens.[10] In the desert of Australia they learned to use rifles, pistols, and knives and practiced hand-to-hand combat. A ship then took them to Port Moresby, on the south coast of New Guinea, and they walked across the island to Finchhaven. "We were crawling on our bellies under fire most of the time," said Cohn.[11] Another ship took them island hopping from Finchhaven to the Philippines. In January 1945 they met with the troops under the command of General MacArthur, who were involved in the battle of Luzon, and entered Manila.

The liberation of the Philippines lasted until July 1945. Once Manila was secured, the U.S. Army moved north to Aparri to prepare for the invasion of Japan in the fall. The men of the 26th Medical Company were in Aparri in August when they heard the wondrous news; Japan had capitulated. The troops were elated: the war was over, and they were going home. They eventually heard what had brought this on. A few days earlier, two powerful "atomic" bombs had been dropped on Hiroshima and Nagasaki. Soon, however, Lane and Mel learned that the war was not over for them. They had been declared "essential" and were shipped to Japan. After the devastation caused by the A-bombs, there was no knowing what diseases and concomitant infections might be encountered. Moreover, Mel's background in physics also singled him out for this duty.

They landed in Hiroshima in the fall. Still in their summer uniforms, the freezing men established a camp just outside the devastated city. Surviving civilians wandered around the camp, and Mel remembers feeding milk to cute little Japanese kids. Although they were famished, the kids hated milk because it was not part of their culture. They would eventually drink it, holding their noses in disgust. The Americans badly needed interpreters, and a young high school English teacher showed up at the camp to volunteer. He had survived the A-bomb only because he happened to be out of town, and upon his return he found that his home and family had been annihilated. The teacher, Kengo Horibata, turned out to offer invaluable assistance to the 26th Medical Company, and Kengo and Mel became friends.[12]

The army tested the environment for radioactivity and microorganisms and then wrote its report. Conventional bombs had essentially destroyed Tokyo and the main hospital was in Kyoto, a city that had been spared. The members of the 26th Medical Company still remaining in Japan were sent to Kyoto. They were now part of an American army of occupation that was mostly idle. This caused disciplinary problems, which required a stronger presence of the military police (MP). Cohn, who had been promoted to noncommissioned officer with the grade of Technical Sergeant, had another interesting experience: he became an MP. It is unclear why he was selected for that post, but by now he had learned not to ask. His job included patrolling the red light district. He also had to learn all there was to know about the procedures for military executions, such as the fact that some offenses committed by army personnel were punishable by death by hanging!

Mel was saved from such a horrific task by an epidemic of diphtheria that devastated Japan. Lane got him out of the military police and reintegrated Mel into his medical unit to help diagnose the disease. Mel traveled all over Japan to attempt to determine the extent of the epidemic,

FIGURE 3. Melvin Cohn in Hiroshima, 1945. Cohn private collection.

collect samples, and work on identifying the bacterium responsible for the illness. It turned out to be an extraordinary strain of *Corynebacterium diphtheriae* that had hemolytic activity. As this was the first original piece of biological research by Melvin Cohn, his report read like a scientific paper.[13] Just as the army job was getting interesting, he was shipped back to the United States on a ten-day "cruise" from Yokohama to Seattle, where he arrived on May 7, 1946.

Hitchhiking across the country was a heartwarming experience for Mel. As he stood by the side of the road in his rumpled uniform, all of the truck drivers knew exactly where he was coming from, and they stopped to give him rides and food. Thanks to the GI Bill of Rights, he was now eligible for tuition and a small stipend, and he could afford to pursue a Ph.D.[14] Since he had majored in physics in college, the choice was obvious. He made appointments with several schools offering a Ph.D. in physics, and he remembers visiting Cornell and Columbia. However, because he had been awarded his bachelor's degree in 1940, and so much had happened in the field of physics during the war, not one physics Ph.D. program accepted him, veteran or not. Dejected, he was sitting on the steps outside Columbia to consider what to do next when he heard a voice calling, "Mel! Mel Cohn! What are you doing here?" It was Captain Pappenheimer. They had not been in touch since they had landed on the south coast of Australia in 1944.

He explained his sad situation to Pap, who told him to forget physics and come get his Ph.D. in protein biochemistry with him at NYU. Pap, who had moved to NYU only shortly before he left for the Pacific, had recently returned, and he had not yet had the time to set up his lab. Mel's first job as a graduate student was to repaint Pap's run-down lab in the Microbiology Department, which was then housed in an old NYU Medical School building. This new department, which had been created by MacLeod, was outstanding. The famous but still controversial Avery-MacLeod-McCarty experiment had been published by then, leaving little doubt that genes were made of DNA. Importantly, the department also included Mark Adams as a faculty member.

Mel enrolled in graduate classes at NYU, where he took Mark Adams's phage course. Pap also sent him to Columbia to learn immunochemistry from Elvin Kabat and Michael Heidelberger. Mel spent summers in Madison, Wisconsin, in the laboratory of Jack Williams.[15] This was where the techniques of ultracentrifugation and electrophoresis had first been developed in the United States, making the lab the American mecca for the study of biological macromolecules by physical methods. Pap completed Mel's scientific education by sending him to the newly organized Brookhaven Symposia and introducing him to the numerous scientists visiting the lab. Among the visitors were Pap's friends, the Frenchmen Monod and Lwoff. It soon became obvious, at least to Pap, that Mel's next training ground should be Paris: Jacques and Mel needed each other.

Monod's early observations of regulatory phenomena in bacteria were made entirely on the basis of a quantitative analysis of the growth of bacterial cultures.[16] Not until 1948 did Monod and his collaborator Anna-Maria Torriani demonstrate that extracts of bacteria grown in the presence of lactose catalyzed the hydrolysis of lactose. At that point Pap argued, and Monod agreed, that further analysis of the system would require a biochemist trained in the latest techniques of protein purification, and in their characterization by physical chemistry and immunochemistry. Pap happened to have trained a student with the right expertise: Mel Cohn.

As far as Mel was concerned, Pap felt that he needed some sophistication best acquired in a city like Paris with a mentor such as Monod. Mel had seen more of the world than most people his age, yet he was not worldly. Seeing beautiful countries under fire—countries that were starving and devastated—was an education in horror, not in culture. It would be educational for him to see Europe in peacetime—a relative term in this period of Cold War, of course. By June 1949, having duly completed his Ph.D. dealing with the diphtheria toxin-antitoxin reaction, Mel sailed to Le Havre on the French liner *De Grasse*. He left with a two-year National Research Council Merck Fellowship, but he stayed in Paris for five years.[17]

Postwar French labs were primitive, and doing protein purification and radioactive labeling without a cold room or a Geiger counter was a tour de force. Nevertheless, Cohn and Monod did pioneering work that established that induction of an enzyme involved de novo protein synthesis,[18] and this launched beta-galactosidase as the first model system to study the mechanism of gene regulation.[19] François Jacob joined Lwoff and Monod at the Pasteur Institute in 1950.[20] Their lab, which was located in an attic and included only a few permanent scientists and students, was largely ignored by the university. In Monod's group were Germaine Cohen-Bazire and Anna-Maria Torriani;[21] Lwoff's group included his wife, Marguerite, and Elie Wollman, who had spent two years at Caltech in Delbrück's lab. However, the Pasteur group enjoyed a solid reputation abroad, especially in the United States. Thanks to the generous support of American foundations, in particular the Rockefeller Foundation, the lab saw a constant turnover of professors on sabbatical, postdoctoral fellows, and visitors.

FIGURE 4. Anna-Maria Torriani and Melvin Cohn on the banks of the Seine in Paris, 1950. Photo by Luigi Gorini, Courtesy Anna-Maria Torriani-Gorini.

During the five years that he spent at Pasteur, Mel met an extraordinary collection of scientists that reads like a who's who of modern biology. It included Seymour Cohen, Mike Doudoroff, Roger Stanier, Louis Siminovitch, Martin Pollock, Dave Hogness, Seymour Benzer, Gunther Stent, Aaron Novick, Sol Spiegelman, and, of course, Max Delbrück. Spiegelman was a regular seminar speaker at Pasteur and a competitor in studying enzyme induction. Jacob wrote a lively account of one of Spiegelman's *corrida* performances when Sol gave a seminar in the attic and was attacked by Monod, Cohn, and Pollock.[22]

The Cold War affected the Pasteur lab, as some American scientists came to Paris to escape the pressures of McCarthyism. This was also the era of the Cold War rhetoric of Trofim Lyssenko, the Russian geneticist who denounced classical genetics and Darwinism as fascist and capitalistic. The Pasteur group's defense of genetics added to their notoriety and attracted leftist senior scientists such as J.B.S. Haldane, J.D. Bernal, and B.C.J.G. Knight. All of these contacts broadened Mel's education and made him a member of a vast network of scientists that had a great impact on his career and on the genesis of the Salk Institute. As far as sophistication was concerned, Mel returned to the United States speaking fluent French—with a Brooklyn accent—and having changed his favorite diet from hamburgers and Coca-Cola to cheese and wine.

In 1953 Mel was offered a position as assistant professor in the Department of Microbiology at Washington University School of Medicine in St. Louis, Missouri. What attracted him there was the fact that, as had been the case at NYU, the department was to be re-created with a brand-new faculty. Arthur Kornberg had just become chairman of the old Department of Bacteriology and Immunology, which he renamed the Department of Microbiology, as MacLeod had done at NYU in 1941. However, Kornberg was a biochemist, not a bacteriologist, immunologist, or microbiologist. Three of Kornberg's postdocs eventually joined him: Osamu Hayaishi, Paul Berg, and, later, Robert Lehman but all were biochemists as well. To fulfill the teaching obligations of his medical school's Department of Microbiology, Kornberg needed faculty with expertise in other areas. Mel Cohn joined them, fresh from Paris, in 1954. He was also a biochemist, but he had trained in microbiology and immunology. In his World War II clinical lab he had dealt with not only exotic tropical diseases but also a whole range of more familiar diseases, including the venereal diseases and wound infections so prevalent in an army. His unusual background was most appropriate for teaching medical students. Perhaps more importantly, he helped attract to St. Louis his former collaborator at Pasteur, David Hogness, and also Dale Kaiser, who had been working on bacteriophages with Lwoff. In this way, with three of the six junior appointments originating from the Monod/Lwoff school, Washington University became a small settlement of the Pasteur Institute.

In spite of the Pasteur émigrés, St. Louis was not Paris. However, between Madison, Chicago, Urbana, Ann Arbor, Bloomington, Lafayette, St. Louis, and Nashville, the Midwest had a remarkable concentration of outstanding and highly interactive biologists in the mid-1950s. Szilard, who had a somewhat uncomfortable appointment in Chicago (see chapter 4), was buzzing around encouraging the exchange of information between the various groups. He attended small informal meetings that were held regularly, mostly in Urbana. It was in those Midwest meetings that Szilard first met two of the future founders of the Salk Institute, Lennox and Cohn.[23] Salva Luria's presence in Urbana made the participants at those meetings something of an extension of the phage group (see chapter 4). The biochemists Irwin Gunsalus and Sol Spiegelman were also in Salva's department. Spiegelman, who had known Cohn very well in Paris, invited him to give a talk in Urbana. This is how Cohn and Lennox met.

Edwin (Ed) S. Lennox was born in 1920, in Savannah, Georgia, to a Jewish immigrant family.[24] Although a well-read man, Ed's father made a living as a peddler. He died when Ed was four years old, after which his mother and her three sons lived with an aunt who ran a kosher restaurant in the Jewish section of Savannah. During the Great Depression, his mother worked as a clerk employed by the Works Progress Administration, which was created in 1935 by Franklin D. Roosevelt to generate jobs. After finishing high school, Ed won a scholarship to enter a new junior college in Savannah. There a professor of history who had received his Ph.D. at Vanderbilt University in Nashville, Tennessee, helped Lennox obtain a scholarship to Vanderbilt, where he arrived in the fall of 1940. One of Ed's instructors in physics turned out to be Max Delbrück. This lucky circumstance was a crucial event in Ed's life.

This was the time when Max was not only teaching physics but was already working with phages (see chapter 4). That was Ed's first contact with a real lab, where people did research and asked questions. Max nurtured him, sent him to take a genetics course, and shared with Ed his interests in music and literature. Ed also met some early members of the phage group who visited Max, in particular Salva Luria. When the war started in 1941, Ed had a deferment to go to graduate school. After his graduation from Vanderbilt in 1942, Max chose Victor Weisskopf in the Physics Department of the University of Rochester to be Ed's Ph.D. advisor. In June, Ed arrived in New York with a fellowship and he got a job as well, maintaining the cyclotron.

In 1942 several senior faculty members of the Physics Department, including Weisskopf, left to join the war effort, and by the summer of 1943 the department had closed. Ed had not yet started his thesis and was still doing graduate course work. Eventually Ed received a letter from Weisskopf indicating that he had a job for Ed in New Mexico. Although the letter contained no details, Ed thought that, since he had lived in the flatlands all his life, it would be interesting to see New Mexico. He accepted the offer without knowing where he was going or what the job was about. He then received instructions to go to Chicago, where someone would put him on a train to Lamy, New Mexico. Once he arrived in Lamy, someone else met him, took him to a little office in Santa Fe, and drove him up the hill to the center on Los Alamos Mesa where the atomic bomb was being built.[25] Ed does not remember the date, but it must have been in early 1944.

At Los Alamos he was assigned, as a civilian, to the Theoretical Division headed by Hans Bethe. The division included several groups, and among the group leaders were Victor Weisskopf, Richard Feynman, and Edward Teller. Soon four junior physics students arrived from Harvard University. All four—Roy Glauber, Kenneth Case, Ted Hall, and Fred de Hoffmann—were about to get drafted. Whether a physics student got drafted or not was probably decided by the local draft boards and was mostly a matter of luck. There was an obvious disparity at Los Alamos between the civilians and the drafted members of the Special Engineering Detachment (SED). This created bitterness, because the SEDs lived in military barracks and were subjected to military rules. The civilians lived in bachelor dorms and lived by one rule only: work. Both Glauber and Case became well-known professors of physics,[26] Hall became an atomic spy,[27] and de Hoffmann eventually became president of General Atomic and later of the Salk Institute (see chapters 5 and 12).

Ed's impressions of de Hoffmann at the time included two interesting observations. First, Fred was closely associated with Teller and did not behave as a regular member of the group. Instead, he hung around with the generals and the powerful administrators who had contacts with the military. The second observation is one that everybody commented on: Fred wore pajamas with a crest of nobility embroidered on them, which seemed somewhat out of place in the shared latrines. There was an active social life on the mesa, and groups with common hobbies formed, from the walkers, the horsemen, and the fishermen to the skiers. This is where Ed became a walker, an activity that became vital for him and eventually fostered important contacts.

After Ed had completed his assigned calculations, Weisskopf sent him with a number of people to the site where the first atomic bomb would be tested. There they put down cables, installed various pieces of recording equipment, and witnessed the Trinity test of July 16, 1945. Meanwhile, they knew that already two bombs had been taken to the island of Tinian and that some location was going to be bombed on the first clear day. At this point they just waited. Then, on August 7, the destruction of Hiroshima was announced on the loudspeakers of Los Alamos. This precipitated a chain of historical events: on August 8 Stalin declared war on Japan, on August 9 a second atomic bomb was dropped on Nagasaki, and on August 14, Japan surrendered. Since

Germany had surrendered in May, World War II was over. The stage was set for "peacetime"— the occupation, the Cold War, and reconstruction.

Meanwhile, at Los Alamos, everyone was preparing to return to civilian life. The junior staff were busy writing reports of their work and taking the excellent classes given by the senior physicists. The many lectures were offered to help the young physicists return to universities, and the National Science Foundation provided fellowships to go back to school. The senior physicists were being actively recruited, and Weisskopf was offered a job that he could not refuse at MIT. Bethe was returning to his department at Cornell University in Ithaca, New York, where he took Lennox as a graduate student. Ed graduated in 1948 with a Ph.D. in theoretical physics and became an assistant professor of physics at the University of Michigan in Ann Arbor. In 1951 and 1952, while teaching at Ann Arbor, he had a summer job in physics at the University of Illinois in Urbana. This is where he became interested in biology. Many biologists visited and gave talks in Urbana, and Ed started doing phage experiments. He got reacquainted with Luria, whom he had met earlier through Delbrück at Vanderbilt. Eventually Luria sponsored him for a postdoctoral fellowship from the National Foundation for Infantile Paralysis (NFIP). By the summer of 1953, Ed started working with Salva. His summertime work was in Cold Spring Harbor, where he reestablished contact with Delbrück. After the summer, he settled in Urbana, where he stayed until 1960.

This was a productive time. Ed did work on photo reactivation of UV-inactivated phages and on transduction; he also made many contacts.[28] A turning point in Ed's career and in the history of the Salk Institute was a visit by Mel Cohn to Urbana to give a talk at the invitation of Sol Spiegelman. Mel had continued to work on enzyme induction while at Washington University, but he had also become interested in the regulation of antibody synthesis. Although induction of an enzyme and induction of an antibody turned out to have very different mechanisms, Mel was tempted to explore both. Mel's training in immunochemistry during his Ph.D. studies with Pap was about to come in handy. The seminar that Ed remembers is one in which Cohn talked about antibodies. Intrigued by the large range of specificities of immunoglobulins, Ed spent the next summer in Mel's lab in St. Louis learning about immunology.[29] They discussed a research project to analyze antibodies made by single cells and initiated it the following summer. By then, Kengo Horibata, the Japanese English teacher Mel had met in

Hiroshima, had arrived in St. Louis. This led to a collaboration that lasted several years.[30]

In 1956 Ed's NFIP fellowship ended and he accepted a position at Urbana in the new biochemistry department headed by Gunsalus. He knew "absolutely nothing about biochemistry," by his own account, and he was hired first as an instructor as he learned biochemistry, chemistry, and genetics. He eventually was promoted to assistant then associate professor of biochemistry while he continued his long-distance collaboration with Cohn. Since Urbana was less than two hundred miles from St. Louis, this was quite feasible, although the logistics were a little complicated. In 1960 Lennox moved to New York, where he had accepted a position as associate professor in the Department of Microbiology at NYU, the same department that had been created by MacLeod in 1941. Pappenheimer, who had succeeded MacLeod as chairman, had moved to Harvard in 1958. The experiments in collaboration with Cohn were completed by 1959, and Mel left St. Louis for Palo Alto, where Jonas Salk visited him that summer (see chapter 5).[31]

The move from Washington University to Stanford Medical School turned out to be one that was not entirely happy for Cohn. The problem was neither the place nor the people: Palo Alto was great, and the colleagues that moved with him were his friends. The Four Musketeers—Berg, Cohn, Hogness, and Kaiser—leased four adjacent lots of Stanford University land on which they built four identical houses, so they became close neighbors as well. The difficulty was that Arthur Kornberg's department developed into one that was focused purely on biochemistry. Kornberg loved enzymes, he was especially fascinated with polymerization reactions, and the focus of the department narrowed further to nucleic acids. Proteins that do not catalyze a chemical reaction, such as immunoglobulins, were of little interest to him. Furthermore, working with intact cells instead of purified proteins seemed out of place. Cohn ended up lecturing on hormones while teaching about immunoglobulins was done by the Immunology Department. Kornberg and the other members of his department were more at ease and happier in a biochemistry department than was Mel, who was made to feel that he no longer fit in. It was time to consider alternatives.

Harvard University was developing a new biology department, where, by 1958, Jim Watson was an associate professor. Since early 1959, before his move to Stanford, Cohn had been approached to accept a professorship at Harvard Biolabs, but he had not seriously considered it because at the time he wanted to remain with the Kornberg group.[32]

Pappenheimer, who had moved to Harvard in 1958, strongly encouraged him to come to Boston. Eventually, in early 1960, Cohn received a firm offer from the chairman of the Department of Biology, Carroll M. Williams, followed by another letter of encouragement from Pap.[33] This was around the same time, March 1960, when Cohn visited La Jolla and ran into Matt Meselson and Jim Watson (who did his best to recruit both Matt and Mel to Harvard; see chapter 5). An amusing letter from Pap says, "I hope you don't go to La Hoya [sic] (there have been rumors)."[34] In view of the developments at Stanford, Mel did visit Harvard and meet Dean McGeorge Bundy, for whom, he found out, he had little affinity. After another visit to "La Hoya" in May and the positive vote from the citizens of San Diego in June, Salk's possibility appeared more solid and tempting. By June, however, Mel received official confirmation of another offer: his friends Roger Stanier and Mike Doudoroff wanted him in Berkeley.[35] Overwhelmed with teaching and with choices to make, Cohn turned to his most trusted friend and confidant, Jacques Monod.[36]

In March 1960 Mel wrote to Jacques explaining what was going on at Stanford and stating his options: Harvard, Berkeley, or perhaps the molecular biology institute planned by Jonas Salk in La Jolla. This was the first time Cohn mentioned Salk's plans, "a great secret," and also that he would like to arrange for a one-year leave beginning in 1961 to spend at the Pasteur Institute. Monod answered right away that he understood and was delighted to hear that Mel might spend a year in his lab.[37] After the land vote Cohn wrote again to Monod, revealing the names of the other people involved at the time: Benzer, Dulbecco, Kalckar, Lennox, Meselson, and Puck. He also asked Monod to meet with Salk, writing, "I am sure that you will like Salk very much," adding, "I have great confidence in his honesty and worldliness."[38] Monod was on his boat sailing somewhere around Corsica at the time and missed meeting with Salk. In August 1960, however, when he returned from his vacation, Monod answered Cohn and brought up the fact that he had received a letter from Lennox asking if he would have space in his lab for him in 1961–62.[39] Monod's longtime friendship with Cohn certainly influenced him in wanting to help. In October 1960, Monod visited Cohn in Palo Alto and learned all about the "secret" institute. Together they wrote the justification and budget of an NSF grant application to cover supplies and equipment for Cohn to work in Paris.[40]

In 1961 Cohn and Lennox both arrived at the Pasteur Institute, having resigned their academic positions, and obtained senior postdoctoral fellowships from the National Science Foundation to work in the Monod lab while assisting in the development of the institute planned by Salk. In June both had received a letter of intent from Salk stating that if and when the Institute for Biology at San Diego materialized, they would be offered positions as Senior Fellows. It was a letter of intent only since, in the absence of a board of trustees, no formal offer could be made. Meanwhile they were in limbo, and Monod played a major role as a founder of the Salk Institute by giving Cohn and Lennox the hospitality of his lab at the Pasteur Institute. Monod, however, was himself interested in the creation of a molecular biology institute. Already in 1958 Monod had sent to Cohn a text he had written concerning a European molecular biology institute.[41] He later was involved in discussions to plan molecular biology institutes at the Sorbonne and at the Pasteur Institute.[42]

Another of Monod's early and most important contributions to the future institute was to attract Francis Crick. Crick could not remember when he first met Monod, but Jacques was so impressed and seduced by Francis upon his first visit to the Pasteur Institute in spring 1955 that he reported that visit to Cohn.[43] Francis and Jacques developed a friendship based on mutual admiration and a shared taste for science and sailing.[44] Monod's deep involvement with the future institute was a strong sign of approval that Crick trusted. In this way, a nucleus of the founding faculty of the future Salk Institute was constituted that held its meetings at the Pasteur Institute. From the start it had an international composition and an attractive location: Paris.

CHAPTER 7

The Spirit of Paris

In the very early days "Bruno" Bronowski and I would fly
from London to Paris to consult with Jonas Salk, Jacques
Monod, Mel Cohn, and Ed Lennox on such fascinating
topics as the by-laws for the proposed institute.

—Francis Crick[1]

While Lennox and Cohn were preparing for their stay in Paris, Jonas
Salk certainly was not idle. The year 1960 had been critical to the creation of the institute, as it was during this time that La Jolla was chosen
as its location and the institute's articles of incorporation and first
bylaws were formulated (see chapter 5). However, the institute did not
yet exist. A few potential Fellows had become interested in the venture,
which some, such as Jim Watson, described as "Jonas' utopia."[2] Indeed,
without a building or a faculty, and nothing more than big plans and
promises, serious candidates were hard to find. There was not even yet
an elected board of trustees, as the first three trustees were the signers of
the articles of incorporation: Harlow. J. Heneman, a management consultant; Kenneth Hoffman, a lawyer from the office of O'Connor and
Farber; and Jonas Salk. After the gift of land seemed assured, securing
trustees was the most urgent task at hand, since only trustees could
appoint faculty. In April 1960, Salk paid his first visit to Warren
Weaver.[3]

Warren Weaver is the godfather of molecular biology. In 1938 he was
first to name the new discipline, which he inspired and supported when
it was in its infancy.[4] Weaver had been the director of the Division of
Natural Sciences at the Rockefeller Foundation since 1932.[5] By then, at
age thirty-eight, he had already had a successful career as a mathematician and was chairman of the Department of Mathematics at the University of Wisconsin in Madison, his alma mater. Born in Wisconsin, and

with a family that lived in Madison, Weaver was bound to attend the University of Wisconsin, where he was trained as a civil engineer and mathematician. He had outstanding teachers at the university, but the mentor who had the greatest influence on his future was Max Mason, a theoretical physicist who was a great teacher and scientist, a leading member of the Madison faculty, and, eventually, a longtime colleague and dear friend of Weaver. Mason left Madison in 1925 to become president of the University of Chicago. After less than three years in Chicago, Mason joined the staff of the Rockefeller Foundation in New York as the first director of its programs in natural sciences, and later he became the foundation's president. It is Mason who eventually convinced Weaver to join him at the Rockefeller Foundation.

Weaver graduated from the University of Wisconsin in 1917, and by September of that year he had moved to Pasadena to take a teaching position at the Throop College of Technology, an institution that would soon be renamed the California Institute of Technology. When Weaver arrived at Throop/Caltech, where there were only four graduate students, the faculty was so small that its meetings were held in the modest office of the university's president. The walls of that office, however, were covered with drawings by a renowned architect, Bertram C. Goodhue, that represented a dream for a spectacular future. Weaver's stay in Pasadena was to be a short one. In 1920 his beloved alma mater called him back, and he returned to Madison, where he stayed for another twelve years. His primary reason for returning to the University of Wisconsin was the opportunity to collaborate with Mason. He enjoyed working with Mason until 1925, when Mason became president of the University of Chicago before joining the Rockefeller Foundation in New York. After Mason left Wisconsin, Weaver completed the classic textbook that they coauthored.[6] Then, in the fall of 1931, Mason called Weaver to New York to convince him to join the staff of the Rockefeller Foundation. This was not an easy decision; Weaver loved teaching and academia, he was doing very well at the university, and Madison was his home. However, as Weaver describes in his remarkably modest autobiography, he believed that, at age thirty-eight, he had probably achieved everything he could as a mathematics professor. Philanthropy, however, opened new possibilities, and, he writes, "We began to pack."

In 1932 Weaver succeeded Mason as director of natural sciences at the Rockefeller Foundation, while Mason became the organization's president. At the time the term "natural sciences" referred to every scientific field other than medicine. Until then, the Rockefeller Foundation had put most of its funding into the physical sciences. Weaver convinced the foundation to shift its natural sciences programs to strongly support quantitative biology. The physical sciences had provided a battery of techniques and tools that would revolutionize the life sciences, and Weaver saw the importance for the foundation to move in that direction.

Weaver spent only three months at the foundation's home office in New York, where he learned the organization's procedures, before he was transferred to their other principal office in Paris. Beginning in the 1920s, the European staff of the foundation constantly traveled to evaluate the performance—and needs—of research laboratories and to discover young talent. This established a system of exchange based on the Rockefeller international science fellowships. These fellowships allowed young American biologists such as H. J. Muller and George Beadle to work in Europe for one or two years while Europeans, including Max Delbrück, Boris Ephrussi, Jacques Monod, and André Lwoff, spent time in the United States. Exchanges within Europe were also encouraged, which allowed, for example, Nikolaï Timoféeff-Ressovsky to move from Moscow to the Kaiser Wilhelm Brain Research Institute in Berlin-Buch, an institute supported by the Rockefeller Foundation (see chapter 4). In the 1930s the Rockefeller Foundation aided the relocation of victims of the Nazi purges in Germany. After World War II, the foundation provided grants to the Cold Spring Harbor Laboratories to support their seminal summer symposia on quantitative biology and helped rebuild the European laboratories. After attending the 1946 Cold Spring Harbor symposium (see chapter 6), Jacques Monod enjoyed an American shopping spree for lab equipment at the expense of the Rockefeller Foundation. He returned to France with some two hundred kilograms of luggage, writing to his wife, "Nous allons pouvoir acheter beaucoup de choses avec les fonds Rockefeller et les doigts me démangent à l'idée des belles experiences que l'on va pouvoir faire avec les nouveaux appareils."[7] Weaver deserves much credit for the revolution that changed biology after World War II.

At the very beginning of World War II, Weaver volunteered to join the Office of Scientific Research and Development (OSRD) as a statistician to design several types of devices that improved the efficiency of antiaircraft fire and protected London (see Prologue). Early in 1941 he was a member of an official scientific mission to London, appointed by Franklin D. Roosevelt to establish contacts with British scientists and military experts. He sailed on a decommissioned old wreck to Lisbon and from there flew to England, where he experienced a terrifying night with an antiaircraft battery outside London.[8] By 1942, having completed the design of an efficient gun director, he was ready for a new task and became chairman of an OSRD agency called the Applied Mathematics Panel (AMP), organized to provide assistance wherever mathematical expertise was needed. It turns out that there were many demands for help from the AMP from the various branches of the military.[9]

After the war Weaver resumed his activities at the Rockefeller Foundation. Amusingly, it was during this postwar period at the foundation, after he had supposedly reached the ceiling of his career as a mathematics professor, that he produced his best-known feat of mathematics instruction. In the late 1940s the president of the Rockefeller Foundation was Chester Barnard, former president of the Bell Telephone Company of New Jersey. One day at lunch he asked Weaver if he had read in the *Bell System Technical Journal* a paper by Claude E. Shannon entitled "A Mathematical Theory of Communication."[10] When Weaver said that he had, Barnard inquired if he had understood the paper. After he again replied "yes," Barnard inquired whether Shannon's ideas could be explained in simpler terms. Yes, again. "All right," said Barnard, "do so." This is how Weaver described the origin of the publication of a small book that became a best seller and a classic.[11] The 1949 book republished Shannon's 1948 paper with a small—but significant—change to the title, which was now "*The* Mathematical Theory of Communication." The addition of an introduction by Weaver made the theory accessible to nonmathematicians and helped give birth to what is known today as information theory.

Weaver was not only a great communicator of science and mathematics but also a scholar of the humanities. He was an expert on *Alice in Wonderland,* and he was a serious collector of rare editions and translations of the book, which inspired him to write the scholarly book *Alice in Many Tongues.*[12] Charles Lutwidge Dodgson, better known as Lewis Carroll, was a mathematician—"not a very good mathematician," wrote Weaver, who was nevertheless fascinated by the tricky and witty mathematical puzzles, strange games and words, puns, and imaginative logic arguments invented by Carroll.[13] Weaver's interest in a children's book may seem curious, but how this affected his interest in the Salk Institute is even curiouser.

By 1952 Weaver was vice-president of the Rockefeller Foundation, and in 1954 he was elected to the board of trustees of the Sloan-Kettering Institute for Cancer Research. It is while sitting on that board that he developed a relationship with Alfred P. Sloan, who appreciated Weaver's frankness and his ability to interpret scientific issues for a lay board. Sloan invited Weaver to join the staff of the Sloan Foundation, but Weaver, ever loyal to the Rockefeller Foundation, did not want to leave it until he had to retire in 1959 (the Sloan Foundation did not have a formal retirement age). In 1956 Weaver became a trustee of the Sloan Foundation, and in 1960 he moved to the Sloan Foundation as vice-president. That was his position when Jonas Salk walked into his office in New York in April 1960.

It appears that this first contact between Salk and Weaver did not go well. Weaver asked probing questions about Salk's plan for the new institute and concluded, "When he left I had a hunch that I was unlikely to see him again."[14] Weaver did, however, follow up that visit with a letter raising several issues that had come up during the course of their conversation.[15] As a teacher, Weaver was concerned that a pure research institute could increase the gap between research and education. He did not consider directing postdoctoral research as teaching, and he hoped that the institute would accept candidates for a Ph.D. degree, if not undergraduates. Moreover, Weaver feared that making the terms of employment at the institute unusually glamorous could lure outstanding scientists and teachers away from universities. In return, Weaver received a grateful letter from Salk, who proposed another visit in late April.[16] The several drafts that Salk wrote of this lengthy letter attest to the fact that he labored to articulate arguments to convince Weaver to join the institute's board of trustees.[17] Meanwhile, Salk wanted to continue their dialogue, as he still hoped that Weaver would eventually be willing to contribute his wisdom and advice to his undertaking. However, Salk stopped short of mentioning that Weaver might become a member of the institute's board of trustees: Warren Weaver was not yet ready to be reeled in.

The National Foundation–March of Dimes (NF-MOD) was not idle at this time either. In an official announcement in the April 15, 1960, issue of the magazine *Science,* it reported that a basic research institute was to be established in California, probably in San Diego, depending on a favorable vote in June on the allocation of city land for the institute.[18] No name was provided for the institute, which was to be headed by Jonas E. Salk. The article quoted excerpts of a draft of what would become part of the institute's articles of incorporation, describing one of the goals of the institute as the advancement of knowledge "relevant to the health and well-being of man, primarily through research in fundamental biology, and in the cause, prevention, and cure of disease, and in the factors and circumstances conducive to the fulfillment of man's biological potential." The new institute was to be generously supported by the National Foundation to the tune of one million dollars a year for operations and at least another million a year for an endowment. The source of funds for the building was not mentioned. Detailed plans, including the names of the trustees and members and architectural plans, were to be announced if the June 7 vote was favorable. Although the vote was, in fact, favorable, a detailed announcement did not appear until two years later, in July 1962.[19]

What happened during those two years has remained largely untold until now, some fifty years later. This two-year period of our institute's history involved many well-known people, strong characters, big egos, much uncertainty and risk, high drama, great dreams, and greater problems. One month after the positive June 7 vote by the citizens of San Diego, on July 9, Salk flew to England for a busy European tour: he planned to visit London, Cambridge, Copenhagen, Stockholm, and perhaps Paris. With a spectacular piece of La Jolla real estate in his pocket, he felt he had something solid to offer: a superb location. He was ready for serious networking and recruiting. He had sent letters announcing his visit to a number of people he wanted to meet, among them Francis Crick, Jacob Bronowski (known as "Bruno" for short), and C. P. Snow. Salk had met Jim Watson in La Jolla in March 1960 (see chapters 5 and 6). Watson was clearly not interested in joining the new institute, only in luring some of the attractive candidates to the Biolabs at Harvard. Salk decided to try to interest Francis Crick instead. One of his first visits in England was to Cambridge to have lunch with Francis. By 1960 Crick was already a celebrity, although he was not yet a Nobel laureate.[20]

Francis Crick was born in 1916 in the village of Weston Favell, near Northampton, a large town in the East Midlands, some seventy miles north of London.[21] The city was famous for its shoe-making industry, and Francis's grandfather directed the family boot and shoe factory, Crick & Co. Francis became interested in science at a young age and won a scholarship to Mill Hill School, a respected boarding school in North London. In 1932 he took the Higher School Certificate exams, which got him accepted at University College London, where he graduated in classical physics in 1939. He had barely started his Ph.D. work when World War II began, the laboratory closed, and a land mine destroyed his thesis project apparatus. Supposedly this was no great loss since, according to Crick himself, his work was on a very boring subject: the effect of pressure and temperature on the viscosity of water. During the war he was assigned as a civilian to work on magnetic mines for the Royal Navy (see Prologue). His wartime service was extremely important to his eventual success, not only because he gained self-confidence and an interest in new topics, but also because he met his mentor Harrie Massey. Massey, born and educated in Melbourne, Australia, was a brilliant and influential mathematician and physicist who obtained his Ph.D. under Ernest Rutherford at the Cavendish Laboratory in Cambridge. By 1940, at age thirty-two, Massey held the chair of mathematics at University College London and had become a Fellow of the Royal Society. It was the support and powerful connections of Massey that opened to Crick the doors of Cambridge University.

At first Crick studied some biology at Cambridge, but in 1949 he started a second Ph.D. project, this time in the unit of Max Perutz, which was supported by the Medical Research Council (MRC) at the Cavendish. Crick, who by then was thirty-three years old, started work on the structure of proteins using X-ray crystallography. This is where he learned the use of models in space to explore the conformation of molecules. This is also where James D. Watson arrived in 1951 to work with John Kendrew. Watson, who was only twenty-three, had completed his Ph.D. with Salva Luria in Indiana (see chapter 4) and spent a postdoctoral year in Copenhagen. Both Watson and Crick were interested in the structure of DNA, and they started the collaboration that would lead to their Nobel Prize. Eventually Crick did obtain his Ph.D., but not for his work on DNA: he submitted his thesis on X-ray studies of polypeptides and proteins in July 1953. However, he did not miss the opportunity to attach to his dissertation reprints of the two DNA papers that had appeared earlier that year.[22]

The day after meeting with Crick, Salk had lunch in London with Bronowski (Bruno for short). Jonas and Bruno had never met before, and the first time that Bronowski is mentioned in the Jonas Salk Papers is on June 6, the day before the San Diego vote.[23] On that date Salk sent a letter to Adriano Buzzati-Traverso thanking him for their recent talk and for sending him a copy of Bronowski's book, which Salk said he "enjoyed tremendously." In addition, Salk wanted to pursue conversations with Buzzati "about the possibilities of an inter-relationship between the institutes in which we are interested."

Buzzati, a population geneticist, was head of the Marine Genetics Division at the Scripps Institute of Oceanography in La Jolla from 1953 to 1962.[24] This is where Salk must have met him, in the spring of 1960, on the eve of the land ballot. Jonas and Adriano soon found out that they had a major aspiration in common: both wanted to create an outstanding institute of molecular biology, Salk above the Pacific Ocean and Buzzati above the spectacular Gulf of Naples.[25] They came up with the idea that their two institutes could start a network of international laboratories of fundamental biology. That concept, and the contact with Bronowski, helped renew and hold the interest of Warren Weaver, whom Salk still wanted to attract to the institute, he hoped as one of its first trustees.

It appears that Weaver and Bronowski had also never met until their involvement with Jonas Salk and his institute. Weaver does not mention Bronowski in his autobiography before meeting Salk in 1960.[26] Ten years later, however, when Weaver's autobiography appeared, he described Bronowski as "the almost incredible British mathematician, logician, philosopher, essayist, dramatist, and poet, the author of a number of superb books."[27] By then he knew Bruno very well, of course, and he obviously had developed a great admiration for him. Weaver's affinity for Bronowski is no surprise. Like Weaver, Bruno was a mathematician, a great communicator, a popularizer of science, and a literary scholar as well. Curiously, while Lewis Carroll fascinated Weaver, Bronowski discovered William Blake, who, like Carroll, wrote imaginative poetry.

Jacob Bronowski was born in Lodz in 1908 to a modest family that lived in Germany during the First World War while Russia occupied Poland. They moved to England in 1920. Bruno believed that the fact that he had to change the language he spoke twice before becoming a writer had a profound effect on him. In London he attended the Central Foundation Boys School, and then he won a scholarship to study mathematics at Jesus College in Cambridge. He completed his Ph.D. in 1933 after writing a dissertation in algebraic geometry, and he was a lecturer at the University College of Hull from 1934 to 1942. During World War II he worked for the United Kingdom's Department of Home Security, where he developed statistical approaches to bombing efficiency. In November 1945 he was deputized to the British Chiefs of Staff Mission to Japan to write a report on the effects of the atomic bombs on Hiroshima and Nagasaki. After the war he did not return to academic life but instead worked for the British government. First he applied statistical methods to industry and economy, and then, in 1950, he joined the National Coal Board as director of research to develop a smokeless solid fuel. This was his position when Jonas Salk visited him in London in the summer of 1960.

By then, while holding his day jobs, Bruno had written poetry and published well-received books about William Blake and other poets, as well as books that introduced science to the layperson.[28] Bronowski was a household name in the United Kingdom, less because of his books than as a result of his performances on the BBC television show *The Brains Trust*. Originally a popular radio quiz show, the show was revived in 1955 as a television series that was watched by millions when it aired every Sunday at teatime. During the show a half dozen questions that had been sent in by viewers were given to a panel of five unprepared personalities for discussion. Bronowski was a frequent panelist, often teamed up with the biologist and author Sir Julian S. Huxley and the philosopher Sir Alfred Jules Ayer. Bruno was a showman, and many did not appreciate his theatrical style; his emphatic slow speech, his affected mannerisms, and his impressive vocabulary could seem pompous and arrogant. Moreover, the show itself was sometimes deprecated as an intellectual game—as entertainment—rather than a genuine attempt to educate the public. As a TV star Bruno was a well-known public personality, but he was not especially respected by other academics. When the chair of history and philosophy of science at University College London became vacant in the mid-1950s, Bruno was not seriously considered for the position. Perhaps he was too public, too successful, and had too many areas of interest. There is also little doubt that envy, as well as a dislike of his style, played a part in hampering his career.

America, however, welcomed him as a lecturer. In the spring of 1953 Bruno was invited as Carnegie Visiting Professor of History at MIT, where he delivered three lectures.[29] Eventually, in 1958, these lectures were published as a book under the title *Science and Human Values*. The earliest paperback edition was published in 1959, the year Jonas Salk first visited La Jolla.[30] It is that now-classic 95-cent little book that Salk received from Buzzati. Bruno's book is said to have initiated, or at least rekindled, the debates over what C. P. Snow labeled the "two cultures." Since Snow did become, in name, a trustee of the Salk Institute, it is relevant to briefly remember who he was.

Charles Percy Snow, widely identified by his initials, was also more formally known as Sir Charles (he was knighted) and eventually as Lord Snow when, in 1964, he took a life peerage and became spokesman for the Ministry of Technology in the House of Lords. Born into a lower-middle-class family, he studied chemistry and physics at Leicester University College, a small provincial institution in his home-town. Through sheer hard work he obtained scholarships to enter Christ College as a Cambridge Ph.D. student in 1928. He started a career as a scientist at the Cavendish Laboratory, but his need to retract a widely publicized early discovery essentially ended his promising career as a scientific researcher. However, his training as a scientist later gave him credibility in the "two cultures" debate. He fell back on another occupation that proved highly successful: writing novels, including the popular series *Strangers and Brothers*. During World War II he was drafted into the civil service, and in 1945 he became a civil service commissioner and industry advisor. By 1959 the success of his novels allowed him to give up regular employment and become a lecturer. The first of his formal addresses, "The Two Cultures and the Scientific Revolution," delivered as a Rede Lecture at Cambridge, remains his best-known lecture by far.

The "two cultures" that Snow opposed were those of literary intel-lectuals and natural scientists. The debate centered on the gap of distrust and incomprehension between the two groups. Snow and Bronowski, however, were entirely at ease in both worlds because both could right-fully claim, "By training I was a scientist; by vocation I was a writer."[31] Arriving in London on July 10, 1960, Salk had made appointments to meet both men. Since Snow was out of town at the time, Salk met Bronowski first. They discussed the "two cultures" and the idea of using biology as a link between the humanities and the sciences.[32] Since biol-ogy raises many ethical, philosophical, and social issues, using an insti-tute for biological studies to build bridges between science and the humanities seemed natural and sound. Jonas and Bruno talked about waiting until the right person came along before developing that idea at the institute. However, judging by the positive way Bruno reacted, Jonas felt, "It may well be that the right person has come along." A lunch meeting with Bruno one week later confirmed that impression. Bruno was very interested in getting involved with the institute. Right after that lunch with Bruno, Jonas met with C.P. Snow and got the idea that Sir Charles, who was then at the peak of his fame but not yet a lord, would make an impressive institute trustee.

Jonas did not know how perfect the timing was, as Bruno was already thinking about leaving his job at the National Coal Board.[33] There might have been internal conflicts at the organization that put Bruno's job in jeopardy. However, one can also easily imagine reasons that making a solid smokeless fuel might have become obsolete, such as changes in Britain's energy policy or clean air standards. In the Atoms for Peace mood of the mid-1950s, one could foresee the burgeoning role of nuclear energy and consider that Britain should stay ahead of nuclear development through investments in atomic reactors rather than in smokeless solid fuel.[34] Moreover, Bruno may well have become bored with his briquettes and seen an opportunity for an attractive career change. Basil O'Connor, who was traveling with Salk on this European tour, had attended the first meeting with Bronowski. His presence must have made the opportunity seem even more serious. Immediately after his return from England, Salk had Lorraine send a copy of Bronowski's book to each of the prospective Fellows.[35] Further contact between Salk and Bronowski soon revealed that Bruno was eager to play some administrative role at the institute in addition to pursuing his creative work. Eventually Bruno, after visiting La Jolla with his wife, Rita, in May 1961, was made a NF-MOD consultant to act as liaison between the scattered prospective Fellows. By fall of 1961 most of the future Fellows were clustered in Europe: Crick and Bronowski were in Britain, while Monod, Lennox, and Cohn were in Paris. Lennox had arrived in Paris in August 1961 and was well settled in with his family when I arrived in September.

I had met Lennox in 1960 at NYU, where he was an associate professor in the Microbiology Department and I had spent a six-month training period in that department while on leave from my job in Brussels. I am a second-generation descendant of the school of Jean Brachet, one of the founders of molecular biology. He and his group at the Université Libre de Bruxelles (ULB) did pioneering work that demonstrated the role of RNA in the synthesis of proteins and contributed much to the concept of messenger RNA.[36] With his international reputation and contacts, Brachet was able to arrange access to the best laboratories for his students. His first graduate student, Jean-Marie Wiame, did postdoctoral work in the laboratory of Carl Cori at Washington University in St. Louis and took the famous microbiology summer course given by Cornelius Van Niel in Pacific Grove. After returning to Brussels, while doing research in bacterial physiology, Wiame started teaching at the ULB a microbiology course inspired by van Niel, and he hired me as an assistant. By 1960 I had a number of publications and connections and was able to obtain a leave of absence to work on gene regulation at NYU in the Microbiology Department. This was the same department that had been founded by Colin MacLeod in 1941, although by then it had been relocated to a new building (see chapter 6). Pappenheimer, who had succeeded MacLeod as chairman, had moved to Harvard, and Bernard Horecker had succeeded him by the time I arrived, so I did not meet Pap then, only years later. However, at NYU I met not only Ed Lennox but also Lane Barksdale, another important character in this story and a member of the Microbiology Department (see chapter 6). Little did I know that some fifty years later I would write a book in which Ed and Lane would play major roles. Their common friend, Mel Cohn, I would not meet until he arrived at the Pasteur Institute in early October 1961.

On October 6, Jonas Salk walked into Monod's lab at the Pasteur Institute and Mel Cohn introduced him to Monod's secretary, Madeleine Brunerie.[37] This was to be a brief visit, when Salk would consult with the three prospective Fellows working in Paris while on his way to London, where he would meet again with Bronowski and Snow, among others. As a result of this shuttle diplomacy, Salk was able to announce by mid-October that a European office had been established for the institute in London.[38] The headquarters of the London office would be the home of Dr. Bronowski, who would keep in touch with the European contingent involved with the institute and assist them as needed. At the same time Salk also invited Sir Charles to serve as an institute trustee and Monod and Crick to become non-full-time Fellows.[39] He mentioned that a special fellowship would also be offered to Leo Szilard.

Clearly, things were happening. With Monod and Crick involved, Lennox and Cohn in Paris, and Salk and Bronowski ready to cross the Channel at any time to assist, the project was picking up momentum. Already in June three new trustees had been added to the three incorporators-trustees already on the board: David Lloyd, a lawyer from Washington, D.C., Melvin Glasser, a dean at Brandeis University and trustee of the National Foundation, and Basil O'Connor, the National Foundation president himself. The fact that Sir Charles had been approached to serve as a trustee as well was an indication that soon a board could be in place that would be large enough to make formal appointments. Decisions would have to be made, and, before signing on the dotted line, it was time to spell out the kind of institute that was being conceived.

This is when the "Spirit of Paris" materialized!

The Institute for Biology would be a new and special kind of place, where ideas could be generated and exchanged. Its atmosphere would cultivate a community of scholars who wanted the freedom to think, to use their imagination, to satisfy their curiosity, scholars who would "be afraid of no issue."[40] It would be a research center with an educational role, whose goal would be to spread ideas to the younger generation and to the public. Concerned very broadly with problems of man, the institute's scholars would approach all questions using the sound scientific methods of the natural sciences, doing research primarily in fundamental biology and its connection with medicine. However, this "freedom to think" would require great flexibility. The scholars would be allowed to change their field of inquiry, to build a group of collaborators of any size and composition—large or small, with senior or junior members—or to function with no group at all. They could do elaborate experiments or no experiments at all, focusing only on theoretical work. They would be able to work anywhere outside the institute and to invite outside collaborators to join them at the institute, for either short or extended periods of time, as they deemed necessary.

This goal placed extreme demands on the structure of our institute. There could be no departments that would limit research to a specific field and teaching to a defined curriculum. The institute's structure—or lack thereof—would be in contrast with that of universities, at which interdisciplinary work was discouraged in favor of a rigid separation of disciplines under the control of department heads. However, an intellectual affiliation with a university was essential, and the new University of California campus in La Jolla was promising. From the beginning it was oriented toward research and science, and it was likely to attract a population of students with critical and eager minds, the kind of students who challenge teachers and make teaching enriching and worthwhile. Our scholars would be free to teach at any level through arrangements with universities such as UC.

However, our institute was not to become a degree-granting institution with the administrative burden that this implies. Indeed, bureaucracy was to be avoided by keeping the size of the institute small and manageable, never to exceed perhaps thirty Fellows and a total staff of about four hundred.[41] A very limited faculty in residence would be enlarged by the addition of a nonresident faculty that would play an important role without requiring a large commitment of space or money. However, the nonresident faculty members would also have to be a special kind of scholar. Interested in ideas, they would look forward to spending time at the institute and find their visits stimulating and rewarding. They would be expected to be in residence for a period of time each year, and occasionally for an extended period, such as a sabbatical year, a summer, or retreat, to think or write, for example.

They would be offered office and laboratory space and could bring students or staff. It would be especially important for the nonresident faculty to spend considerable time at the institute because the they would have power. They were not to be mere advisors, but, to avoid inbreeding, they would have a vote in faculty appointments and promotions. It was therefore essential that they get to know the people and the place personally. They would also facilitate communication and exchanges with a network of institutions around the world.

As recommended by Oppenheimer during the meetings in Pittsburgh in 1958, the institute would be set up as a private corporation, with its own articles of incorporation, bylaws, and board of trustees (see chapter 2). The Fellows, resident or not, were to be the creative members of that corporation. Therefore, they should be scientist-citizens and share the duty—and the privilege—of voting on important issues. There was to be a democratic process. Although the Fellows would be well aware that financial decisions rested in the hands of the trustees, the Fellows would have a degree of control over academic decisions. As the trustees, who would nominate and elect new trustees, would also ratify the appointment of faculty, the Fellows who nominate new Fellows could veto the appointment of trustees. However, the two groups, trustees and Fellows, should cooperate and have a chance to discuss and understand each other's positions. What better way to achieve this than by allowing trustees and Fellows to be on both sides of the fence? Some Fellows would also be trustees and some trustees could be elected Fellows.

This was certainly an unusual structure for an academic institution and would require a serious revision of the bylaws. The "Parisian" Fellows called back Salk and Bronowski for a discussion in Paris on October 22, 1961. On that Sunday, Salk, Bronowski, Lennox, and Cohn met at Monod's apartment. It is at that meeting that amendments to the bylaws were proposed that would embody the "Spirit of Paris." As it turned out, discussions of the institute's bylaws became so involved and passionate that Jonas, making light of touchy issues, invented a new academic discipline: "Bylawlogy!" Eventually, a proposal for revised bylaws was drafted.[42] Small, informal, and flexible, our institute would be more of a community than an institution, almost like a family. In fact, a residence for Salk, the institute's director, was planned in the meeting center, but that was never built. Our small community of scientist-citizens would be enlarged by a constant turnover of visitors, and quarters for temporary visitors were included in the original building plans (see chapter 8).

The terms of resident Fellows' appointments were easy to agree upon, if not to implement. They did not want to be called professors, not because they did not want to teach, but because they wanted to convey the fact that teaching was not their primary function. Their salaries would be high but not excessively generous, comparable to that of a full professor at Harvard. A more unusual and attractive aspect of the offer was "estate" funds provided by the institute to cover expenses unforeseen or not allowed in grant budgets. The most important condition, however, was that the founding resident Fellows' appointments would be "without term," meaning for life. This practice was modeled after the Institute for Advanced Study at Princeton, where the original faculty consisted of permanent members. Oppenheimer and Salk entirely agreed on this point, and the archives reveal that this was the first condition spelled out in Monod's handwriting at that Sunday meeting in his apartment.[43] This was the major difference in terms between the Salk proposal and all other job offers. This point was especially important for the members of the Greatest Generation. Having started their Ph.D. work or postdoctoral training after their return from service during World War II, most were in their thirties before they obtained their first academic position. Crick and Szilard in particular were concerned about this, because they would qualify for only very small pensions upon mandatory retirement.[44] Both eventually became resident Fellows of the institute and remained productive until their last day. Indeed, having a job for life was also a great attraction for Monod. Crick recalled that in spring 1976, when he visited Monod in Paris, they made plans to meet at the institute and perhaps write a book together after Monod had retired as director of the Pasteur Institute.[45] However, this was not to be, as Jacques died suddenly only a few weeks later.

In December 1961 the Fellows in Europe received the welcome news that the institute's board of trustees had been increased from six to twenty-one members, a group that included Bronowski, Snow, Weaver, Gerard Piel, and other prominent personalities such as Edward Murrow.[46] In the late 1940s, Piel, a friend of Warren Weaver, had bought the old-fashioned magazine *Scientific American* with borrowed money, and, with the help of Weaver, he eventually transformed it into the attractive and informed publication that we know today. Jonas Salk had written an article about the polio vaccine that had appeared in the magazine in April 1955, the month of the announcement in Ann Arbor, and Piel wanted to help. It was Piel who convinced Weaver to join the board of the institute, a task made easier because Weaver was impressed by the Fellows that Salk had been able to recruit, by the inclusion of Snow and Bronowski among the trustees, and by the institute's principles of freedom, flexibility, and small size. Weaver was elected chairman of the board, and this gave our institute some instant credibility that it badly needed.

The first officially recorded Fellows meeting was held in Paris on February 4, 1962.[47] The "Parisian" Fellows discussed the scientists whom they wished to attract to the institute, and they contacted a number of them informally.

By March 1962 the trustees authorized the director to invite eight people to accept resident Fellow appointments: Seymour Benzer, Paul Berg, Jacob Bronowski, Melvin Cohn, Renato Dulbecco, Edwin Lennox, Matthew Meselson, and Theodore Puck. Invitations to four potential nonresident Fellows—Crick, Monod, Szilard, and Weaver— were approved as well. The inclusion of Weaver among the potential nonresident Fellows created an unexpected crisis. By May, before these invitations were to come up for a vote at a Fellows meeting, Crick was furious and threatened to resign! Indeed, Weaver had not been nominated through the proper channels—that is, by the Fellows. In a letter to Monod, Crick questioned the qualifications of Weaver as a Fellow, even a nonresident one, asking, "When did he last publish a scientific paper?"[48] Eventually proper channels were used, and at the Fellows meeting of May 19, 1962, the four nonresident Fellows were unanimously elected.[49] The same meeting would complete the crystallization of the body of Fellows. Berg and Meselson declined invitations at that time, but six invitations were accepted, although Benzer and Puck eventually resigned before joining the institute. This left a group of only four Fellows (in addition to Salk) who were ready to move: Bronowski, Cohn, Dulbecco, and Lennox.

On June 1, 1962, two years after the 1960 ballot vote of the citizens of San Diego, the organization of the Institute for Biological Studies was announced in the magazine *Science*.[50] One might wonder why it took two years to recruit a handful of trustees and Fellows, and why there was still no building for the institute to move into. Also, why was Gerard Piel mentioned in the article as the president of the institute, while Jonas Salk was only its director, and why had Theodore Puck already been dropped from the list of Fellows? Much had been going on behind the scenes that had not been revealed to the press or even to the Fellows themselves. It all had to do with a simple and sad truth: the NF-MOD was essentially bankrupt!

CHAPTER 8

Our Dear Kahn Building

We shape our buildings, and afterwards our buildings shape us.
—Winston Churchill[1]

The first public announcement of the NF-MOD's financial troubles appeared in various newspapers, including the *New York Times,* on July 1, 1960, less than one month after the June 7 vote. Charles Massey, who was then national director of MOD chapters, candidly discussed with the press the foundation's problems.[2] The Salk vaccine had been so successful that public concern about polio had decreased dramatically, reducing contributions to the organization by more than 50 percent since in 1954, the year before the Salk vaccine became available. Meanwhile, the foundation was still paying for the care of polio patients for whom the vaccine came too late, and in 1955 it had bought enough doses of the vaccine to inoculate nine million children free of charge. The foundation was also in the process of redefining its target diseases, which were to include birth defects and arthritis. Worthy as these causes are, they did not generate the fear—and generosity—caused by the polio epidemics. Eventually, by deferring debts and changing some of its policies, the NF-MOD made a remarkable recovery, but this took some years.[3] Meanwhile, in the early 1960s, when Basil O'Connor made big commitments on behalf of the NF-MOD to support Jonas Salk's dream institute, the foundation was in deep trouble. It is therefore not surprising that the NF-MOD did not commit to providing funds for the new institute's building, only for operations and endowment (see chapter 5). It was understood that building funds were to be sought from private donors or other foundations.[4]

Jonas, however, did not seem overly concerned. He thought of O'Connor as a twentieth-century Lorenzo de' Medici, the famous Florentine patron of artists and scholars.[5] After the development of the polio vaccine, Salk believed that there was no limit to what Doc could accomplish—and spend. Salk was probably counting on Doc's power and contacts to raise more funds for his institute, either through the MOD or from other sources. Meanwhile, Jonas had his own priorities. After the pleasant shock of discovering La Jolla in August 1959, he made his second visit in November with O'Connor to make sure that Doc approved of the site (see chapter 5). It soon became clear that it was critical to select an architect in order to close the land deal. As luck would have it, in the fall of 1959 Louis (Lou) Kahn gave a lecture entitled "Order in Science and Art" at the Carnegie Institute of Technology in Pittsburgh. A colleague mentioned the lecture to Jonas, as well as the fact that Kahn was completing a laboratory building for the University of Pennsylvania in Philadelphia, the Richards Medical Research Laboratories. Salk contacted Kahn and arranged to visit him in December to see the Richards lab site and hopefully to receive expert advice about the selection of an architect. That visit marked the beginning of the friendship and collaboration between Salk and Kahn.[6]

The decision to establish the institute at the La Jolla site was made in January 1960, after Salk met with the San Diego city fathers.[7] The city council was to meet in March to hear statements by Salk and O'Connor to support the proposal of a gift of land for the institute. An architectural site model was urgently needed to illustrate the proposal, and Kahn was able to prepare one in time for the city council's consideration. However, Kahn's first scheme for the laboratory building, which was based on the Richards building, was quickly abandoned after the presentation to the city council. Although Kahn had visited the La Jolla site with Salk in February and was well aware of the layout and size of the land, Kahn did not carefully think through the model, perhaps because of the rush to prepare it. He proposed clusters of towers that were appropriate on the crowded University of Pennsylvania campus but were lost on the spacious La Jolla site. For that and other reasons, the original plan was quite unsuitable, but it was attractive enough to be approved by the city, and the proposal of the gift was disclosed to the public.[8] After the favorable June 7 vote, Kahn began working on a low-rise arrangement. Although the NF-MOD expected that funds for construction would be raised from other sources, it had advanced some money for the organization and development of the institute, including the preparation of architectural plans.

Meanwhile, and still without a cent in his pocket for the building, Jonas left on July 10 for his European tour (see chapter 7). He certainly knew about the difficulties faced by the NF-MOD, since the story had appeared in the July 1 newspapers. However, he probably felt that the best thing he could do to persuade potential donors was to secure a stellar faculty and prestigious trustees by giving wide and wonderful publicity to the future institute. He made numerous and impressive contacts, and his notes read like a who's who in science and philosophy.[9] Those contacted included a number of Nobel laureates (Archibald Hill, Hugo Theorell, Peter Medawar), future laureates (Francis Crick, Rita Levi-Montalcini, André Lwoff), and polymaths such as Sir Isaiah Berlin, Michael Polanyi, and, of course, C. P. Snow and Jacob Bronowski. Salk also told Lew Thomas, professor of medicine at NYU, all about the institute and concluded, "He can be counted among the good friends." One might wonder if Jonas was perhaps a little naïve and if he ever realized that, by bragging about his dream institute, he also made good enemies. However, his European tour was successful insofar as it enticed Bronowski to join the project, and 1960 ended on a positive note with the incorporation of the Institute for Biology at Torrey Pines (see chapter 5). Still, no building funds had yet materialized.

By February 1961 Jonas must have been desperate. He wrote a surprising letter to his friend and confidant Ted Puck, who was also a prospective Fellow.[10] Salk suddenly wanted to return to the original idea of setting up his institute in Pittsburgh (see chapter 2)! Clearly the major consideration for this change of heart was that "the matter of financing appropriate facilities for the Institute would be resolved." Operations could begin promptly, if temporarily, in vacant space of the Municipal Hospital. Salk then listed a number of obvious cultural advantages of Pittsburgh over San Diego: the presence of Carnegie Tech, the Carnegie Library and museums, and the Pittsburgh Symphony Orchestra; the availability of clinical research facilities; its proximity to major East Coast cities; and the beauty of the area around Pittsburgh— "to say nothing of the city itself." Nothing happened to reverse the La Jolla plan, but by the end of April Salk had informed O'Connor that the idea of the institute would have to be abandoned. Jonas "had tried and had failed to raise even $1 for construction."[11] Since there was to be no institute, Jonas was about to cable Bruno to cancel his May visit to the United States to meet the people involved and see the La Jolla site.

However, "Mr. O'Connor's determination was strong." He proposed to raise the funds for construction through the mechanism of the NF-MOD on one condition—that Salk allowed the institute to bear his name. Doc insisted that he could not raise the money otherwise and left Jonas little choice: Salk Institute or no institute. Salk knew then that he was in trouble. Just as had been the case with the media hype surrounding the polio vaccine, Salk was caught between O'Connor's need for publicity to raise funds and the opinion of the scientific community (see chapter 1). Jonas was concerned that the use of his name would appear self-serving and self-aggrandizing. Naming an institute after a living scientist was considered inappropriate, and Jonas would suffer some abuse for it, but he had to agree. The name change took place very discreetly in December 1961. It seemed important to keep it secret as long as possible, especially as the body of Fellows was shaping up in Paris (see chapter 7). The revelation to the prospective Fellows that their institute was to be called The Salk Institute could very well turn some of them away from the entire venture. Jonas was afraid that this might happen, and he was right. If the announcement could be delayed, the name change would be a fait accompli and funds for the building would be available as a result. At least that is what he hoped!

On December 1 a special meeting of the institute trustees took place in O'Connor's office.[12] The institute's board of trustees then consisted of only six members (see chapter 7), and the four of them present at that special meeting constituted a quorum. At the request of the city of San Diego, the name of the institute had already been changed once, from the Institute for Biology at Torrey Pines to the Institute for Biology at San Diego. On December 1, O'Connor, Salk, Heneman, and Glasser voted that the name be changed to the Salk Institute for Biological Studies, San Diego, California. Immediately following that vote, the number of trustees was increased from six to twenty-one, and seven of the additional fifteen trustees were elected right away, including Bronowski, Snow, Piel, and Warren Weaver, who would serve as chairman of the board. Significantly, several bank accounts were opened instantly, and financial arrangements were made that would allow the Salk Institute to come to life using funds provided by the NF-MOD. On December 14 the trustees of the NF-MOD authorized a fund drive to raise fifteen million dollars for the organization and construction of the Salk Institute.[13] Ending 1961 on a happy note, on December 19 a deed of land was granted by the city to The Salk Institute for Biological Studies, San Diego, California.

The names of the newly elected trustees were quickly communicated to the prospective Fellows and were well received, as expected (see chapter 7). It appears, however, that the new name of the institute was not mentioned at the same time. The name change actually required changing the articles of incorporation, and that would take some time to complete before the new name became official. In all of the 1962 correspondence between Salk and the Fellows, the institute was simply called "The Institute for Biology." Since there was no institute and no institute stationery, all letters were typed on plain paper or on University of Pittsburgh letterhead. There was no rush to inform the Fellows of the embarrassing new name. The Fellows were kept happy. Their amendments to the bylaws had been approved, official invitations for appointments had been extended, and the first Body of Fellows had been established (see chapter 7). One can only wonder how the idealistic demands of the prospective Fellows would have been answered if money had been plentiful.

There was, however, a rush to announce the constitution of the faculty before the start of the fund-raising drive. On May 31, 1962, Gerard Piel held a news conference for science writers in New York City.[14] The press received proofs of an announcement to appear in the magazine *Science* the next day.[15] It was Piel who had invited the press as president of the Institute for Biological Studies that was under construction in San Diego. The point of the announcement was "the most important news there can be" about an enterprise of this kind: the organization of the faculty of the institute. Before introducing Salk as director and Weaver as chairman of the board of trustees, Piel made an unusual statement. He pointed out that the name of the institute used in the magazine article—Institute for Biological Studies—represented the wishes of Jonas Salk but that the board of trustees firmly insisted that it would be called the Salk Institute for Biological Studies. The name in *Science* took into consideration the feelings of Dr. Salk, "whose reluctance on this question you will all appreciate."[16]

At the news conference Salk announced the names and credentials of the resident Fellows and nonresident Fellows.[17] Two future Fellows who were not in Europe, Szilard and Dulbecco, attended this well-publicized event.[18] The unusual organization and governance of the new institute were also mentioned to the press. Eventually O'Connor addressed the question of financing and the role of the NF-MOD and announced the drive to raise fifteen million dollars to build and equip the institute, saying, "We start tomorrow, June 1." There was to be a banquet to launch the drive and a site dedication ceremony on June 2 with a speech by Rabbi Morton Cohn.[19] O'Connor hoped to collect the entire fifteen million dollars during the month of June.

Unfortunately, the NF-MOD fund-raising drive for the Salk Institute building failed miserably. The MOD had previously had remarkable success and experience in raising small sums from millions of people, but to raise fifteen million dollars in one month required at least some large gifts. Perhaps the MOD lacked the necessary tactics or contacts to succeed on such short notice. However, one can imagine more profound reasons for their failure.

In the first place, the timing was disastrous. Seven years had passed since Jonas Salk had become a public hero (see chapter 1), and much had happened since 1955 that had led to criticisms of the National Foundation for Infantile Paralysis (NFIP), which later became the NF-MOD.[20] After the unveiling of the vaccine at Ann Arbor, the foundation was blamed for rushing the licensing of the Salk vaccine, which led to the tragic Cutter incident, which killed eleven people. The countless children saved by the Salk vaccine were easily forgotten, since they had remained healthy and anonymous. The foundation was also seen as too aggressive in its ways, and the remarkable methods of the NF-MOD for fund raising were suddenly criticized as being too successful, as mobilizing too much money for a rare disease. The virulent attacks by Sabin against the killed vaccine had also done harm to the reputation of both Jonas Salk and the foundation. The powerful American Medical Association (AMA), which had been left out of the 1955 Salk vaccine trials, had become openly critical of the foundation as well. Even the NF-MOD's generosity in providing free polio vaccinations to millions of children was criticized as taking away control—not to mention profits—from the medical profession. The AMA had endorsed the Sabin vaccine for use in the United States even before it was officially licensed in 1962. The early 1960s were the Sabin years; the Sabin vaccine replaced the Salk vaccine for some forty years in the United States, and the World Health Organization recommended the Sabin vaccine for global eradication of polio (see chapter 1).[21] By 1962 the name of neither Salk nor the NF-MOD was as popular as it once had been.

Considering the circumstances in 1962, perhaps it had been a mistake for O'Connor to insist that Salk's name be attached to the institute. Salk also probably made an error by relying on the NF-MOD to raise funds for the building, but he had no experience with fund raising, and he did not know where else to turn. In 1963 a second, more traditional, capital campaign was organized that sought large gifts directly to the Salk Institute, but it raised only about one million dollars. In the summer of 1963, both Warren Weaver and Jonas Salk contacted the president of the Sloan Foundation, Henry Heald, to try to obtain support for the building, but the response was flatly negative.[22] This was not surprising since Alfred Sloan, who was still alive though no longer president of his foundation, simply despised O'Connor. To make matters worse, O'Connor had denigrated not only the Sloan Foundation but also other major scientific foundations, including the Rockefeller Foundation.[23] There was little hope of getting support from major foundations, at least in the early 1960s.

As long as most of the Fellows remained in Europe, it was quite easy to avoid revealing to them details about the institute's financial situation. However, in March 1963 all of the prospective resident Fellows met in San Diego. That meeting "marked a decisive turning point in the history of the Institute and its evolution into a unified scientific community."[24] They had received a gracious invitation from David Bonner, chairman of the Biology Department of the new University of California campus at San Diego (UCSD).[25] Bonner offered them temporary space in new UCSD buildings, which would be ready in the summer but remain partially unoccupied for about a year. This would allow the Fellows to move to La Jolla that summer and would initiate fruitful interactions between the two neighboring institutions. The offer was discussed at the March meeting, where it was decided that the institute would start operations on July 1 in temporary quarters to be built on the institute site. Indeed, the Fellows had gained much confidence by what they saw in La Jolla. On March 14, two weeks before the meeting, the NF-MOD had made funds available to cover organizational, operating, and research expenses and deficits of the institute until 1966. In addition, the NF-MOD had approved a grant of eight million dollars for construction. However, there was a catch: the remaining seven million dollars of the fifteen million dollar estimate was to be raised by the institute itself with help of its trustees and friends. Meanwhile, with funds advanced by the NF-MOD, construction had started and the excavation and foundations were almost completed.

As far as the Fellows knew, "No one seems very worried about the financial affairs."[26] In the fall of 1963, the Fellows started arriving in La Jolla and setting up their labs in temporary quarters, even though funds for the building "were not yet in sight" (see chapter 9).[27] Since no building funds existed and no specific construction budget had been established, there was—at least in Jonas's mind—no limit to what could be spent![28] This was an example of Jonas's talent for viewing a potential disaster as an opportunity. He was not bluffing—he sincerely believed that the financial problems would eventually be resolved—but this was one of those times when to call him overly optimistic would have been an understatement. Jonas was right to assume that some building funds would become available. Nevertheless, the financial crisis of 1962 threatened the creation of the Salk Institute and lingered for decades. It profoundly affected the institute's development, and its consequences are still with us today.

Following the visit of the Richards building in Philadelphia and the gift of land granted by the city in 1960, Salk and Kahn started dreaming. Much has been written about the affinity between the two men, and Salk is said to have been Kahn's favorite client.[29] They listened closely to each other, but not every thought had to be explained because they intuitively understood one another. They were both mystics and intuitive. As a lab scientist, Salk knew what was needed; and, as the director and only faculty member and administrator of a nonexisting institute, he was in a position to make unopposed decisions. Jonas hardly had to explain to Lou that laboratory work was only one aspect of a scientist's activity. This was to be a community of scholars, not just a workplace. It would require everything that scholars might need to satisfy their minds and bodies. Since the community would remain small, it would also be essential to provide the intellectual stimulation brought by visitors, both short- and long-term. Salk did not give Kahn a program or a budget, only a general idea of the kind of community he was dreaming about.

The La Jolla site, both challenging and inspirational, dictated the location of three building groups: laboratory buildings, a meetinghouse, and residences for visitors.[30] A large section of flatland stood at the head of a canyon of parklands. This flatland was an ideal building site for the laboratory as its eastern edge was a public road, making it easily accessible.[31] The building site had two arms that extended west to the sea, along the north and south edges of the canyon. Kahn proposed to locate the meetinghouse and the residences for visitors at the west end of the site, close to the sea, each on one arm of land at opposite sides of the canyon.[32] Sadly, Kahn's project was to remain unfinished for lack of funds.[33] Neither the meetinghouse nor the residences ever materialized, but Kahn's design for the Salk Institute Meeting House will survive as an unbuilt masterpiece.[34]

About a year after Kahn's presentation of the first laboratory plan to the city, he proposed a low-rise version. The first two-story plan consisted of four parallel buildings defining two gardens. Having signed in April 1962 an agreement with contractors to build this version, Salk almost immediately had the plans revised. One of his concerns was that the two separate gardens would split laboratory personnel into two groups. Another was the insufficient flexibility in the design of the laboratories.[35] In May Kahn proposed a new and final version, with two laboratory buildings and a single garden. However, to provide enough laboratory floor space these buildings needed to have three floors. This meant that, in order to satisfy the height limits set by the California Coastal Commission, one floor of the buildings had to be below ground level. This created very enjoyable sunken patios and light wells.

Although the towers had been scrapped, the concept of the "servant" and "served" spaces developed for the Richards building had been preserved. The idea was to house the utilities—mechanical equipment, pipes, and ducts—in a space separate from the laboratory that they served. In the Richards building, the services were located in towers; at the Salk Institute they were given their own floors. This method would prevent problems encountered with the Richards building, but it would also turn out to be very expensive. Indeed, this plan doubled the number of floors, from three to six: each laboratory floor had its own interstitial floor—or pipe space—above it, feeding services down through the ceiling. The buildings were engineered to create unobstructed and flexible laboratory space. This provided the feature essential in Jonas's mind—the flexibility and capacity to evolve in order to accommodate changing scientific technology and staff.

Construction started in June 1962 in spite of the failed NF-MOD fund-raising drive. Thanks to the NF-MOD pledge of an eight-million-dollar grant, construction continued until the summer of 1963, when it became clear that it would be necessary to borrow funds. This is when architectural work on the meetinghouse and residences was halted.[36] In January 1964, after six months of failure to borrow funds, a loan of ten million dollars was negotiated using as collateral the ten-million-dollar endowment pledged by the NF.[37] By July 1965 the entire fifteen million dollars had been spent and construction stopped. The south laboratory building was still an empty shell, without mechanical and electrical services. Construction of the north laboratory building was completed, but it still needed outfitting.

This left us with the north half of an empty laboratory building, the south half of a building shell, no endowment, and the (now-iconic) courtyard a muddy mess. Eventually a number of smaller gifts and a loan from the NF-MOD to match a grant by the federal government allowed the outfitting of the north building. In 1966, after three years in temporary barracks, we moved into the new labs.[38] Dealing with the field of dirt between the two buildings then became pressing. Thanks to a generous trustee, Theodore Gildred Sr., in 1967 the planned garden became a courtyard paved with travertine marble designed by the Mexican architect Luis Barragán.[39] Meanwhile, the resident Fellows obtained research grants from the NIH and other sources to support the direct costs of their research. The money available for indirect costs was insufficient, but the institute survived with the help of the NF-MOD.

Exactly how dear was our Kahn building? After the fifteen million dollars was spent, we simply stopped counting. Over the years, as more brave scientists joined our faculty and more funds became available, the south building was slowly activated. This took decades. Actually, it took until 1996, when a new and controversial east building was completed,[40] to transform the middle floor of the south building into laboratories. One could say that Kahn's south building was never finished, at least to match its north counterpart. People who, like me, have worked in both buildings over the years are well aware of significant differences between the two that tell a sad tale. In the offices and studies of the south building, cheap carpeting replaces the beautiful hardwood floors of the north building, and flimsy shades have taken the place of the heavy teak louvered shutters that were intended.

Fortunately, we did not wait for the south building to be completed to enjoy it. In the absence of a meeting and recreation center, the empty shell was a blessing. By 1969 its middle floor, as unobstructed as a football field and almost as big, held two prefab seminar rooms, one small and one large, their interior walls painted different shades of purple, the favorite color of Rita Bronowski. The space outside the seminar rooms opened onto the new and beautiful travertine courtyard and made a perfect reception area for what became our temporary conference center. Administrative offices and small meeting rooms were built all around the third floor against its exterior glass walls. The top floor became a great recreation area that held gym mats, table tennis, and even a kiln. The bottom floor was an ideal if unnecessarily fancy location for on-site storage, and we still miss it badly.

Exactly how appreciated was our Kahn building is an interesting question. It actually took some time for it to be recognized as a masterpiece by nonarchitects. Our friend André Lwoff, the French scientist from the Pasteur Institute (see chapter 6), used to say, "On dirait une prison,"[41] This reveals that he was sensitive to the architectural style that became known as Brutalism. Interestingly, a sociologist who spent two years at our institute and wrote about life in its laboratories chose to ignore the Kahn building.

How do the people who work in the Kahn building feel about it today? One must admit that most of us have a love-hate relationship with Louis Kahn, depending on the weather. Our building clearly belongs in California. On rainy and windy days the windows facing the ocean leak, water seeps under doors: offices, studies, labs, and pipe spaces get flooded. Gutters are unknown here; they have been replaced by scuppers, openings that allow water to drain from the roof and upper floors directly onto innocent bystanders. Outside stairways become dangerous waterfalls, the courtyard is slippery, and the wind can blow you away. In heavy weather an emergency maintenance crew stays on duty all night to avoid major damage. Don't ask them in the morning how they like the Kahn building.

Yet when the sunshine returns and dries up the marble, much is forgiven. Our iconic courtyard, properly called the Theodore Gildred Court, again welcomes tourists and architects. It certainly is a nice site to visit and take pictures, but it is not a comfortable place for scientists to meet, sit, and talk. The glare, the hot afternoon sun, and the hard marble benches are not friendly. In fact, the courtyard stands empty most of the time. What our institute misses most is a comfortable meeting place for the staff, one of the amenities that the meetinghouse would have provided.

Nonetheless, we love our courtyard and enjoy the shows that it offers. Kahn provided a dramatic stage for watching natural happenings: spectacular sunsets to witness the elusive green flash or the sun setting in line with the courtyard's central canal twice a year on the equinoxes. A bonus that neither Kahn nor Barragán could have predicted was the evolution of the Torrey Pines Gliderport (see chapter 5).[42] Originally established in 1930 as a soaring site for full-scale sailplanes, it became a popular location for hang gliders in the 1970s and paragliders in the 1980s. This has resulted in a constant parade of colorful gliders hovering above the ocean between the two symmetric Kahn buildings. The vision is so surreal it looks like a live Magritte.[43]

Admittedly, the pipe spaces are the guts of the building. Jonas used a more elegant biological term, calling them the "mesenchyme" spaces that connect and feed the labs. They have proven themselves over the years, as the in-house building crew can reconfigure labs almost at will while causing only minor disruption to neighboring areas. The disadvantage of this flexibility is that it is almost too easy to reconfigure the space. Since the Kahn building begs to be remodeled, construction never stops. One can wonder whether all of these changes and renovations are really necessary. This flexibility is probably a recruiting asset, however, as job candidates appreciate being given an empty space in which they can design their own new lab. And some uses of the pipe spaces could not be foreseen and were clearly unintended. In the early days of lax security, homeless people found a cozy refuge there between the warm ducts, and there have been rumors of pipe-space romances.

Clearly Kahn and Salk shaped our building, but to what extent did our building shape us? Our building gives us our identity; it makes us unique and proud. It is our greatest asset, it is beautiful and solid, and it will be there long after we are gone. It gives confidence that our research is equally substantial. Finally, it attracts outstanding faculty and generous donors.[44] Detractors of Jonas Salk and his institute have criticized him for spending an exorbitant amount of money for the building. They claimed that he erected a monument to himself at the expense of research. And it has often been said that to construct such a building, the Salk Institute must have been either well endowed or directed by people who were financially irresponsible. Even Warren Weaver thought that we should have started the institute in a more modest facility.[45] Although the cost of our building did make our beginnings difficult, the patience of the founders paid off and gives us what is our most valuable advantage today. It was a remarkable investment, if not a wise one. It took courage and a vision that was, sadly, aborted before it could be fully realized. To consider Winston Churchill's dilemma, if it were destroyed, would we rebuild it in its present form today? In keeping with the original spirit of the place, that would have to be put to a vote of the faculty and trustees.

Of course, no building existed when Lennox, Cohn, and I arrived in the fall of 1963. The Kahn building site was only a hole in the ground, and construction had stopped while a loan was being negotiated. However, the barracks had been built and were in the process of being equipped. By the end of 1963, the Salk Institute would have a total of thirty-three employees, and we would start our lab work as we were discovering the local community. There is no denying that La Jolla came as quite a cultural shock after Paris. But there was no time to be homesick as we did our first experiments in the lab, set up a new home, and prepared for the first colloquium and Fellows meeting in the barracks.

CHAPTER 9

Pioneering

We are the makers of manners, Kate.

—Shakespeare, *Henry V* (5.2.263)

The barracks into which we moved in 1963—two wooden frame buildings painted white—were quite functional. The first building completed
housed Jonas's laboratory and included facilities to be shared: a tiny
library/seminar room and the laboratory kitchen, where culture media
were prepared and laboratory glassware was washed and sterilized. The
second building was to house the labs of the other resident Fellows. A
deck facing the ocean connected the two barracks. With a few garden
tables and chairs, this was where we sat to have lunch and to talk. The
deck was mostly the domain of Leo Szilard. There he would enjoy the
California sunshine, snooze, and "botch" (i.e., think with his eyes
closed). As people crossed the deck, sometimes he would stop them to
ask surprising questions about what they were doing. If Leo extracted
his rumpled notebook from the pocket of his baggy trousers, you knew
you had his attention.

It was easily agreed that the mild Southern California winters made
February the ideal month for the annual Fellows meetings, which
brought the resident and the nonresident Fellows together in La Jolla
and were followed by a trustee's meeting. In addition, it was decided to
organize an annual scientific colloquium at the same time to explore
various areas in which new Fellows might be appointed.

The first Fellows colloquium took place in 1964 in an unfinished lab
in our barracks. Some of the best scientists in the emerging field of neurobiology were gathered there. It was the very first scientific meeting at

the Salk Institute site, the first neurobiology meeting on Torrey Pines Mesa—a memorable success and an enormous boost to our morale. Although the Fellows agreed that understanding the brain was essential to understanding man and that neurobiology was to be the next frontier in biology, it was Francis Crick who first suggested organizing the meeting. In a letter addressed to Jonas Salk, dated November 25, 1963, Francis mentioned that Jacques Monod had visited him the preceding week and that they had discussed plans for the first Fellows meeting to be held at our institute in February of the following year.[1]

Crick's letter made two suggestions that were to have a great impact on the future of our institute. The first was to consider Leslie Orgel as a Fellow and the second was to organize a colloquium that would include experts on the nervous system. Orgel was going to be at MIT for about six months. He was fielding job offers that required him to make a decision by early March 1964, but he could visit La Jolla in February. Francis insisted that all of the Fellows should be persuaded to adjust their schedule in order to spend at least a week together at the Salk and meet Orgel.

An important issue for discussion by the Fellows in February 1964 was in which areas the institute should make appointments and how to identify suitable candidates. Francis believed that two main fields in biology, embryology and the study of the higher nervous system, were the most important. At the time, neither field was represented at the institute, and the Fellows needed to know more about both subjects. Both Jacques Monod and Mel Cohn knew Clifford Grobstein, who was then at Stanford, and thought well of him as an embryologist. In the area of the nervous system, Francis recommended Roger Sperry and made a strong case for David Hubel of Harvard Medical School.

It is amusing that in his autobiographical book Crick mentions that he first learned of the Hubel and Wiesel studies from a footnote in an article in the literary magazine *Encounter*.[2] By chance, the November 1963 issue of *Scientific American*—which had just appeared at the time of Francis's November 25 letter to Jonas—contained an article by Hubel that Crick described as very exciting.[3] It was based on a paper published earlier in collaboration with Torsten Wiesel and concerned their pioneering studies of the visual system of the cat.[4]

Although Crick's letter mentioned only Orgel, Grobstein, Sperry, and Hubel, it also suggested that Salk sound out the other Fellows. The resulting meeting turned out to be a remarkable one. Crick recalls in his book that the meeting was very small, with only about a dozen speakers and a small audience, but the listeners were such a formidable and critical group that the last speaker was visibly trembling![5] The audience included not only Basil O'Connor and Salk but all eight of the founding Fellows. Also in attendance were the few people working in the labs at the time, including myself and Marguerite Vogt as well as two longtime technicians of Jonas who had moved from Pittsburgh with him, Elsie Ward and Francis Yurochko (the latter manned the slide projector). Szilard's wife, Gertrude Weiss (Trude), and Jonas's oldest son, Peter, attended as well. It was at the same time a prestigious gathering and a family affair. Louis Kahn's spectacular plan for the entire dream project (see chapter 8) was proudly displayed on the wall of the modest barracks.

While Jonas had moved to La Jolla in December 1962 and had the barracks built, Lennox and Cohn arrived in the fall of 1963, and Bronowski in January 1964. Dulbecco came to La Jolla briefly to attend the February 1964 meeting but then returned to Glasgow, where he spent a sabbatical year at the university. It was in February 1964 that I met Renato Dulbecco for the first time. His path to the Salk Institute, which started in Italy, is well worth recalling here.

Renato was born in Catanzaro, a southern city and the capital of Calabria, but he was raised in the native village of his father, the Ligurian town of Porto Maurizio, on the Italian Riviera, only twenty-five miles from the French border.[6] His father was a civil engineer who had been sent to Calabria to help rebuild the region after it was devastated by the Messina earthquake and tsunami of 1908. That is where he met Renato's mother, a Calabrian from Tropea. At the end of the First World War, when Renato was five years old, the Dulbecco family returned to Porto Maurizio to share the modest apartment of his paternal grandparents. Renato's parents taught him to write and count at a very young age. This helped to put him ahead of his classmates in elementary school, and he continued to do very well in high school in Oneglia, a village near Porto Maurizio. It is Mussolini, Il Duce, who had combined the two rival villages, Oneglia and Porto Maurizio, into a new town he called Imperia. The name, deeds, and loud speeches of Mussolini were increasingly in the news when Renato was growing up, but they seemed not to be taken too seriously in Italy. Fascism, however, continued to spread, and by the time Renato was in high school he was wearing the black shirt of the Fascist youths, like all his schoolmates.

When Renato finished high school, in 1930, he had to decide whether he would attend university in Turin or Genoa. His father had studied at Turin, and it was decided that Renato would do the same. He enjoyed studying science, but he was unsure what subject to choose. However, he was curious about medicine, and the fact it involved so many special-ties appealed to Renato. He was sure to find a specialty that suited him. In his first two years of medical school Renato was fascinated by anat-omy, which was taught by Professor Giuseppe Levi, a powerful faculty member who was admired not only for his teaching and research but also for his open opposition to Fascism. Such opposition was unusual at the time in Italy, where most people were naïve about politics and indif-ferent to the consequences of political ideas. Every year the professor took on a few second-year students as interns at the Institute of Anat-omy, where research in biology was carried out. Renato was accepted as an intern of Professor Levi, as was Rita Levi-Montalcini. Having started medical school the same year, Rita and Renato knew each other, and they became better acquainted as they were initiated into experi-mental biological research.

In 1936 Dulbecco completed his medical degree and was called to military service as a physician for an infantry regiment. After two years of service he obtained a position as assistant to Professor Vanzetti, his thesis advisor from the Institute of Pathological Anatomy at his alma mater. In 1940 he was briefly caught in the turmoil of World War II. After invading Belgium in May, the German army had marched into France and reached Paris on June 14. With a German victory in sight, Italy declared war on Britain and France on June 10 and invaded south-ern France on June 21. Dulbecco was sent to the French border, where skirmishes lasted only a couple of days before an armistice was signed. He immediately returned to Turin, but by spring 1942 his regiment had been called to fight alongside the German army in Russia. After a long trip by train they arrived in Ukraine in the heat of the summer. The Russian winter caught up to the soldiers when they reached the frozen River Don and the Russians attacked. As a physician, Dulbecco did not participate directly in the action, but he saw its consequences as scores of wounded Italian and German soldiers were brought to the rear to the infirmary. During this conflict, however, Dulbecco slipped on the ice and took a lucky fall that dislocated his shoulder but probably saved his life.[7] He woke up in an army hospital and was sent back home for a six-month leave. His regiment was wiped out as the Russians conducted their major offensive on the River Don.

Home by March 1943, Dulbecco resumed his work at the university, but, because of the frequent bombardments of big cities by the Allies, he eventually settled in Sommariva Perno, a village some twenty-five miles south of Turin. There he set up a successful private practice, earning a living by taking care of the medical and dental needs of the population. He was due to report back to the army in September of that year, but during his six-month leave so much had happened that returning to war simply made no sense. In July Mussolini had been arrested and the Fascist government had fallen, but not for long. On September 8, just as Dulbecco was due to report back to the army, Italy surrendered to the Allies. However, the Germans then marched into Italy, occupied Turin and Rome, and reinstated Mussolini. One week later the Allies entered Naples, and as they gained ground, Italy declared war on Germany. This state of confusion and opportunistic politics was typical of Italy at the time. From one week to the next, many people in Italy, even the soldiers, did not know who was an ally or an enemy. Dulbecco spent the rest of the war in Sommariva, where at least he could be of some help treating people from the village and partisans who were hiding in the area. As the Allies approached the city, it wasn't until April 1945 that the Germans left Turin without a fight.

After the war Dulbecco resumed his work at the University of Turin, where he had been reappointed as an assistant lecturer at the Institute of Pathological Anatomy. In the summer of 1945 he met again his former classmate Rita Levi-Montalcini, who had a similar position with Professor Levi in the Department of Anatomy at the medical school.[8] Rita and Renato had not seen each other since the beginning of the war, and they had so much to tell. While catching up on what had happened to them during the tragic war years, they became good friends. After the Racial Manifesto of July 1938, Rita, a Jew, had no longer been allowed to attend the university. Professor Levi had been in the same situation and was no longer able to teach. By March 1939, the conditions in Turin had become too dangerous, and Rita accepted an invitation from Professor L. Laruelle to continue her work at the Centre Neurologique in Brussels, Belgium. This opportunity was especially attractive since it would allow her to stay in touch with Professor Levi, who had also moved to Belgium, as he had been invited to work at the University of Liège. After September 3, 1939, when Britain and France declared war on Germany (see the prologue), the invasion of Belgium became imminent and Rita returned to Turin. At first she set up a private lab in her home, where Professor Levi eventually joined her in her work on chick embryos. By 1942, however, the bombings of Turin had become so intense that the little lab had to be moved to a farm in a mountain village. After the German invasion of Italy in 1943, Rita and her family and friends from Turin filled out false identification cards and moved to

Florence, where they were not known as Jews. In September 1944, the Allies liberated Florence, but fighting continued for months in the Apennine Mountains north of town, bringing thousands of refugees into Florence. Facing an epidemic of typhoid fever, Rita became a doctor and a nurse for the Allies' health service until the end of the war. In spring 1945 she was finally able to return home to Turin.

While Rita was pursuing her interesting observations with chick embryos, Renato was unsure of what direction to follow, although he was certain that he wanted to do research. During this turning point in his career, Rita—who was a few years older than Renato—acted as his mentor and gave him wise advice. She suggested that he consider joining the lab of Professor Levi, who had returned to his institute and was looking for an assistant. Upon Rita's recommendation, Renato was accepted and became interested in the development of chick embryos. This led him to observe the mysterious effects of radium radiation on development. This research appeared to be worth pursuing, but Renato knew nothing about radiations and their biological effects. Rita suggested that he take physics classes at the university, which he did for two years while teaching histology. Rita then wrote to Salva Luria to inquire whether he could offer Renato a fellowship to work for a year in his lab at Indiana University in Bloomington (see chapter 4).[9] Rita knew Salva very well, although he had started medical school one year ahead of her. Renato knew him very little, but they had an opportunity to get better acquainted when Salva visited Rita's lab in Turin in 1946. Renato and Salva discovered that they had several interests in common, including the effect of radiation on living cells. Salva invited him to work for a year in Bloomington, and in the fall of 1947 Renato boarded the Polish ship *Sobieski* in Genoa for New York on his way to Indiana University.[10]

In Bloomington, Salva welcomed him like a brother, and Renato discovered a new world, both in the lab and outside. He easily adapted to life in the United States and met some fascinating scientists, including Salva's student Jim Watson, "un étudiant très intelligent, mais qui avait des façons un peu étranges."[11] Salva's research dealt with phages, and Renato took to the subject quickly. In 1948 Salva extended Renato's contract for another year. Salva also invited him to attend scientific meetings, where Renato kept in touch with Rita, and to spend the summer of 1948 working at Cold Spring Harbor, where he met many famous scientists, including Max Delbrück. The main upshot of Renato's move from Turin to Bloomington was that he met Max, who invited him to work in his laboratory at Caltech. In 1949, Renato moved to Pasadena as a senior research fellow in Delbrück's lab.

To bring Dulbecco to Caltech, Delbrück had used funds from the NFIP that were intended to support basic research into the biology of viruses, including bacterial viruses (bacteriophages). However, in October 1949 additional funds were pledged from another source: the James G. Boswell Foundation. A member of the Caltech board of trustees, J. G. Boswell suffered from painful shingles, a viral disease for which there was no known treatment. The $100,000 gift to Caltech was to study the biology of animal viruses and particularly of viruses that cause disease in man.[12] The first use Delbrück made of the Boswell funds was to organize at Caltech, in March 1950, a symposium to assess the state of the art of animal virology. Progress in that field was hindered by the fact that animal viruses were usually grown in live animals, and the use of tissue cultures was in its infancy. However, the director of research at the NFIP, Harry Weaver, was well aware of the recent progress made by Enders and his colleagues, who had been able to grow poliovirus in roller-bottle cultures (see chapter 1). At Weaver's suggestion, Delbrück visited Enders's laboratory in the summer of 1950. He was impressed. After his return to Pasadena in the fall, Delbrück offered Dulbecco the possibility of interrupting his phage work for a few months in order to explore recent techniques to grow animal viruses in tissue cultures. Delbrück wanted Renato to spend time in a number of tissue culture and virus labs, including Enders's. Renato had seen the successful use of small tissue cultures by Professor Levi and Rita in Turin, and he accepted with enthusiasm. In January 1951 he set out for a three-month tour with a check from the Boswell Foundation in his pocket and the blessings of Delbrück and of George Beadle, the head of the Caltech Biology Division.

Dulbecco, having learned much about tissue culture and viruses, was back at Caltech by March convinced that virus research needed quantitative assays and that methodology developed for bacterial viruses could be applied to animal viruses. He proposed to adapt to an equine encephalitis virus a plaque assay that had been developed for bacteriophages, and Delbrück trusted that the plan would work (see chapter 4). By April, lab space was being equipped in a subbasement location as a precaution against possible danger to humans. In January 1952, Beadle reported Dulbecco's first positive results to James Boswell and obtained another $125,000 to fund the development of the animal virus field at Caltech. Dulbecco was immediately promoted to associate professor and reported his results to the National Academy of Sciences.[13] Harry Weaver quickly offered Dulbecco generous support from the NFIP to extend his techniques to poliovirus research. The NFIP provided him with a fully equipped lab off the Caltech campus because of concerns about polio. Caltech supplied him with a collaborator recommended by Delbrück, Marguerite Vogt, and by 1954 the plaque technique had been successfully adapted to poliovirus.[14]

This technique not only provided an improved assay for virus but also allowed the isolation of pure lines of virus. Moreover, different shapes, colors, or sizes of plaques could distinguish mutants. Ironically, it was Dulbecco's plaque assay that made possible the isolation of attenuated strains of poliovirus for the development of the Sabin vaccine. It is poetic justice that Renato became a founder of the Salk Institute. He had been chosen in the Doering-Szilard memo as a tissue-culture specialist (see chapter 4). Jonas had contacted him soon after discovering La Jolla, and Renato had visited the Torrey Pines Mesa site in spring 1960 at the same time as Meselson, Puck, Cohn, and Watson. Renato had not participated in the Paris discussions of 1961–63, but he had accepted the invitation to become a Fellow in May 1962 (see chapter 7). Having resigned from Caltech, he arranged to spend a year at the University of Glasgow while the Salk Institute building was being constructed. By March 1963 Renato had agreed to begin work in a temporary building erected on the site (see chapter 8). This brought him to our barracks in February 1964, where he had the pleasure to meet again his former classmate Rita Levi-Montalcini.

Indeed, Rita was one of the fourteen invited speakers at our February 1964 colloquium Embryology and Differentiation—The Nervous System: Facts and Hypotheses. Other speakers included Ross Adey, Joel Elkes, Clifford Grobstein, Oscar Hechter, David Hubel, Roy John, Seymour Kety, Oscar Lowry, Walle Nauta, Charles M. Pomerat, James David Robertson, Roger Sperry, and Fred Wilt. In addition there were three discussants who did not give a talk: Donald Glaser, Robert Livingston, and Leslie Orgel.

Although the list of participants was certainly impressive, Rita was the speaker who most impressed me personally. Though I had heard much about her from Mel Cohn, who knew her at Washington University in St. Louis in the mid-1950s, when he was an associate professor in the Department of Microbiology, this was the first time I had met her (see chapter 6). This was the time when Rita worked in Saint Louis, invited by Viktor Hamburger. Later she would collaborate with Stanley

FIGURE 5. The First Neuroscience Symposium on Torrey Pines Mesa, February 1964. In the front row, from left to right: Leo Szilard, Francis Crick, Renato Dulbecco. In the second row, from left to right: Charles Pomerat, Rita Levi-Montalcini, Melvin Cohn. Courtesy The Salk Institute.

Cohen on work that led her to the discovery of a nerve growth factor, and, in 1986, to their shared Nobel Prize for their discoveries of growth factors. Rita walked up to the barracks like a queen, looking both classy and sexy on high heels, wearing a tight black skirt and gorgeous antique jewelry. Rita was invited to become our first female non-resident Fellow, but she declined because in the early 1960s she was setting up a new laboratory in Rome. Commuting between the two continents, she felt that she could not spend the two to three weeks in La Jolla expected annually from nonresident Fellows.

The stimulating colloquium talks were constantly interrupted by questions. Hubel, who could stay for only one day, gave a one-hour talk on Sunday that stretched to three hours by public demand. Following the meeting, Jonas received letters of thanks and congratulations that he forwarded to the Fellows.[15] The terms used by the participants to qualify the meeting included "exciting," "the best biology meeting ever attended," "fascinating," "one superb presentation after another," and "a gem." Statements like "The Institute is off to a good start" and "The

FIGURE 6. The First Neuroscience Symposium on Torrey Pines
Mesa, February 1964. From left to right, Rita Levi-Montalcini
and Marguerite Vogt on the deck of the barracks. Courtesy The
Salk Institute.

meeting foreshadows the future of your great enterprise" made us feel
hopeful. We felt that this kind of event had been well worth waiting for,
and watching the participants basking in the California sun on the deck
of our barracks confirmed that February was the right month for our
annual meeting of Fellows.

That first Fellows meeting foretold the future of San Diego's academic institutions: Leslie Orgel joined the institute as a resident Fellow
in September 1964, Bob Livingston eventually founded the Department
of Neuroscience at UCSD, and Cliff Grobstein became dean of the
UCSD Medical School. Unfortunately, due to our financial crisis (see
chapter 8), the institute did not start its neurosciences program for some
time. However, Stephen Kuffler, the boss of Hubel and Wiesel, became
a nonresident Fellow in 1966 and later played an essential role in developing neurobiology at the Salk Institute, but that had to wait until the
early 1970s.[16] Moreover, the addition of Leslie Orgel was critical to the
institute's development and was most fortunate for me, as Leslie became
a key collaborator in my doctoral work.[17]

Shortly after the February 1964 meeting, Leo Szilard became a resident Fellow. Leo had accepted a position as nonresident Fellow in May 1963 with the option to become a resident Fellow on equal footing as the other Fellows at a later date, an agreement that was the result of lengthy negotiations and correspondence between Leo and Jonas. Leo's resident Fellow appointment became effective on April 1, 1964.[18] Finally he had a steady job for life, in sunny California and in a growing intellectual community that included many friends. Leo and Trude had moved to La Jolla in February and lived in a cottage at the Del Charro Motel. After a lifetime of living apart or in hotel rooms, this was the closest they ever came to settling down in a private home. Then came the phone call on Saturday morning, May 30: Leo had died in his sleep that night of a massive heart attack. He was sixty-six. He had been a resident Fellow for two months. With him died some of Jonas's dreams of creating an institute for biological studies with broader humanistic goals and a strong social conscience.

Two weeks later, on June 13, a memorial was held for Leo in the unfinished Kahn building. A striking but depressing site, the building looked like the ruins of a citadel after a war.[19] The towers, still short, looked damaged, as if by a bomb, with gaping openings where doors and windows would have been blown out by the explosion. The holes in the concrete walls looked like the result of gunshots. Bare steel structures loomed high. Crooked slabs of concrete marked the podium and the stage for a string quartet. Metal folding chairs, lined up in rows, remained largely empty. Two friends of Leo reminded us of his influence on the world. Carl Eckart, a geophysicist from the Scripps Institute of Oceanography, talked about Leo's importance to physics. Ruth Adams, who had been editor of the *Bulletin of the Atomic Scientists* and a participant with Leo in the Pugwash Conferences, talked about his role in the cause for peace. Salk and Lennox were the speakers for the institute.

In spite of having our hearts in our throats, we were still able to smile at Leo's wry humor as his version of the Ten Commandments was distributed to those at the memorial. In 1940, Leo had written his own Ten Commandments in the style of Moses. He wrote them in German but was never happy with any translation so they had remained unpublished.[20] Right after Leo's death Bronowski had translated them into English for distribution at the memorial, and eventually that translation was published in a volume of recollections coedited by Trude.[21] Leo was obviously much less strict and funnier than Moses. For example, Leo's

FIGURE 7. Leo Szilard's memorial in the unfinished Kahn building, June 13, 1964. Courtesy The Salk Institute.

seventh commandment sounded more like friendly advice than a command from God: "Do not lie without need." It was also at Leo's memorial that Trude revealed the contents of a note he left with his last wishes for the disposal of his ashes at a small ceremony: put them in a little box tied to balloons and, with a few friends, watch them ascend to heaven.[22] Leo could be blunt—he could wear us down like sandpaper—yet we miss him: his imagination, his humor, his concern for the world, and his complete contempt for worldly possessions. He never wanted to own anything of material value—no house, no car—only well-used books and a couple of suits. True to the rule he adopted during the First World War (see chapter 3), he arrived at the pearly gates carrying all his possessions in two suitcases and in his soul.

Both Szilard and Orgel were interested in the aging process, and each had published a paper on the subject.[23] Szilard had proposed holding a symposium on aging at the institute and was supposed to get involved in its organization. Orgel was the obvious choice to take over that task after he joined us as a resident Fellow in September 1964. Leslie had been introduced to the institute with a letter of recommendation from Francis Crick, who described him as having one of the keenest intellects in molecular biology.[24] However, Orgel had started his career as a theoretical inorganic chemist. He had done such outstanding work in that field in England that he had been elected Fellow of the Royal Society (F.R.S.) in 1962, at age thirty-five. What made Orgel so perfect for our institute was his extraordinary choice of research topic: the origins of life. He was afraid of no issue and used sound science to approach problems; that was what our institute was all about (see chapter 7). Leslie provided the ideal link between theoretical chemistry and biology.

Leslie was born in London in 1927.[25] He was a teenager when World War II started in Europe and too young to be drafted, but he was nevertheless caught in the Blitz: his high school was evacuated to Bedford, north of London, to escape the heavy bombardments.[26] He was awarded a B.A. (1949) and a D.Phil. (1951) in chemistry from Oxford University, where he began his career as a theoretical chemist. His interest in biology really started during his postdoctoral research fellowship at Caltech in 1953–54. Although his fellowship was formally with Linus Pauling, the famous Nobel laureate in chemistry, Leslie spent most of his time with the founders of molecular biology, who gravitated around George Beadle and Max Delbrück. This was an exciting time, as the DNA double-helix structure had just been proposed by Watson and Crick.[27] Molecular biologists were then pondering how the four-nucleotide sequence of DNA could determine the sequence of the twenty amino acids in proteins. At Caltech Leslie became deeply involved with problems concerning the genetic code and the structure and function of RNA. After a six-month stint at the University of Chicago, he returned to England in 1955. Two years later he became the assistant director of research in the Department of Theoretical Chemistry at Cambridge University, and in 1963 he became a reader in Cambridge's chemistry laboratory. His outstanding work in theoretical inorganic chemistry had resulted in numerous publications, and his election as a Fellow of the Royal Society eventually would have eventually won him a prestigious position in England. However, Leslie had become more interested in biology and wanted to change fields. This was essentially impossible within the framework of the British educational system. Leslie commented, "If you started as an inorganic chemist, by and large you stayed an inorganic chemist. Not impossible to move, much more difficult than in America, and very much less difficult in a brand-new institute that didn't yet quite know what it wanted to do—I mean, where it was going to go."[28]

Indeed, the Salk Institute, young and without binding traditions, allowed Leslie to change fields very successfully, and he spent the rest of his career with us, productive until his death in 2007. Steadfast, savvy, and cool, he helped the institute survive many crises and grow. One of his early contributions to the intellectual life of the institute was to organize the program of the 1965 symposium, Topics in the Biology of Aging.[29] This was a pioneering meeting because the field of research on aging was in its infancy: Leslie called it a morass. However, this meeting was interesting in two ways. First, the edited proceedings of the meeting were published as a book that was to be the first of a series of monographs from the Salk Institute. Unfortunately, due to a lack of funds, it was also the last. Second, this book, which was published in 1966, included a paper about aging in cell cultures by Leonard Hayflick, who was a key participant at the meeting.[30] In the early 1960s he had made observations that were very controversial at the time but remain highly significant today. He had obtained evidence that—contrary to accepted belief—normal human cells in tissue culture are not immortal: they have a finite capacity to replicate before they die. He conceived the existence of an as-yet unidentified molecular mechanism able to count cell divisions and involved in cell aging. Hayflick's studies and ideas turned out to be the prelude to today's active field of research on telomeres and their role in cancer and aging.[31]

The symposium on aging took place in November, but much had happened at the institute earlier in 1965. The first important event of the year had been the February Fellows meeting. After the tremendous success of the 1964 neurobiology colloquium, the Fellows had to consider what to do for an encore. Amazingly, the 1965 Fellows colloquium was even more remarkable than our first neurobiology meeting, although for somewhat different reasons. It was a memorable workshop that not only changed the course of immunology but also provided high drama.[32]

The Antibody Workshop was the brainchild of Ed Lennox and Mel Cohn, who had met in 1955 when Cohn, a faculty member at Washington University in St. Louis, had given a seminar on antibodies at the University of Illinois in Urbana (see chapter 6). Antibodies were considered intriguing because they belong to a class of proteins, the immunoglobulins, that have an extraordinary number of specificities in spite of their limited variation of physical properties and amino acid composition. To study antibody formation, Ed and Mel had started a collaboration that lasted several years.[33] However, they soon came to discover communication difficulties among researchers working on antibody formation because this subject attracted scientists with such different interests and backgrounds. Lennox and Cohn saw the need for small and informal meetings to overcome this communication gap and to foster fruitful interactions between the different specialists interested in the chemistry and biology of the immune system. Lennox had been strongly influenced by Delbrück and the phage group. Cohn had combined his training in immunology with the bacterial genetics thinking of the Pasteur group. Ed and Mel had complementary expertise and connections. In 1959 they obtained a National Science Foundation grant to support Antibody Workshops, and by 1965 they had sponsored seven small but highly successful meetings.[34] Aimed at scientists involved in the new field of antibody structure and function, these workshops were casual, like the early phage workshops. The programs were flexible, the talks were not published, and there was ample time for discussions; young investigators and foreign visitors were welcome. These small meetings accelerated progress in immunology by promoting the rapid exchange of unpublished results, ideas, and materials and by encouraging the standardization of procedures. Most important were the lively discussions and exchanges of criticisms and of suggestions for experiments. This open atmosphere is difficult to imagine today in the age of competition and secrecy generated mostly by patenting. In the early 1960s, the number of investigators involved in this field was so small and the subject so obscure that comprehension was more important than competition. However, this changed dramatically after the 1965 Antibody Workshop.

By 1965 the field was ripe enough to be "molecularized." Until then the molecular biologists had shown little interest in immunology, probably bored by the phenomenology and terminology involved. However, it was clear that the amino acid sequence of antibodies would hold the key to understanding their formation and functions. Since DNA sequencing had not yet been developed, painstaking protein sequencing would have to be done. Lennox and Cohn decided to focus the February 1965 Antibody Workshop on protein sequencing and to convince their molecular biologist friends to join the immunologists. To arouse the interest of molecular biologists in the immune system, Cohn invited scientists working not only on the structure and sequence of immunoglobulins but also on the genetics of the system and on its general biology and evolution. Bringing together the pundits of immunology, protein sequencing, and molecular biology would, if successful, mean at least one hundred participants, more than our barracks could possibly hold. Moreover, it was decided that organizing a retreat in an attractive location where the participants could bring their families might entice some nonimmunologists to attend.[35]

One Sunday in November 1964 Mel Cohn and I drove to Warner Springs to assess its suitability.[36] Located in the foothills of Mount Palomar, only an hour away from La Jolla, the Warner Springs Guest Ranch was a resort hotel in the Anza-Borrego desert. It was perfect. Isolated and very informal in the old California ranch style, it had all the amenities necessary plus historical charm.[37] Its conference facilities were excellent, and the location even had a private airport. One month after our visit, Cohn sent a rather funny letter of invitation to some 140 scientists, about a hundred of whom accepted.[38] Mel described the facilities as "a series of adobe cottages which have fireplaces and nice living rooms," where discussions could continue late into the night. He also wrote, "If you are coming with your wife (or equivalent), or other members of your family, tell us so that we can give special accommodations." Indeed, Max Delbrück came with Manny and two babies. Odile Crick accompanied her husband, Francis. Even Warren Weaver brought the Mrs. With sessions in the morning and at night, the afternoons were free. Mel enclosed "some propaganda on Warner Springs" illustrating the recreation facilities, which included two Olympic-size swimming pools, one fed by natural hot springs and the other cold, "which the English will certainly want to swim in at 7:00 in the morning."[39] The group rate was sixteen dollars per day per person, meals included!

All of the resident Fellows except for Dulbecco attended the meeting. The three nonresident Fellows also accepted the invitation, though Monod had to cancel, as he arrived in La Jolla after the meeting. We missed Leo, but it was nice to see his collaborator and longtime friend Aaron Novick among the nonimmunologists in the audience. Other nonimmunologists—in addition to Crick, Delbrück, and Novick—included Chris Anfinsen, Seymour Benzer, Rita Levi-Montalcini, Lubert Styer, and Jim Watson. The list of immunologists who attended reads like a who's who in the field with the exception of Rod Porter, who was unable to come. Also present were a number of future immunologists, including Lee Hood, Paul Knopf, Fritz Melchers, Mike Parkhouse, and Norbert Hilschmann, at the time an unknown German postdoc working in Lyman Craig's lab at the Rockefeller Institute.[40]

The first two days of the meeting were informative but uneventful. The third day was designated for immunoglobulin sequence work expected to consist of preliminary data with little actual sequencing. With hindsight, it is amusing to read in Porter's letter to Cohn, "If you had Hilschmann and Putnam it might be dull—sequence work usually is—but it would be valuable to have the results collected."[41] However, on that Wednesday morning the audience was stunned when Hilschmann showed a slide with the virtually complete amino acid sequence of an antibody chain.[42] This started a commotion as many people reached for pen and paper to copy the sequence. The commotion changed to loud protests when, after only a few seconds, Hilschmann requested that the slide be turned off. The session chairman was Jon Singer, who remained remarkably calm in the midst of what Doolittle described as "something very close to pandemonium."[43] Singer managed to restore a measure of order and tried to reason with Hilschmann, who was obviously worried about being scooped by his powerful competitors sitting in the audience. The whole incident was both exciting and sad to watch. Only two months after this dramatic workshop, Hilschmann and Craig submitted a paper reporting the now-famous sequences.[44] Their paper presented the first evidence that an antibody chain consisted of a constant and a variable region, products of two separate genes joined, somehow, as a single polypeptide. Hilschmann's result had opened a new field of inquiry in which the molecular biologists became very interested indeed. The elucidation of the molecular mechanism involved and its role in generating antibody diversity kept immunologists and molecular biologists busy and arguing for years.

This marked the end of immunology as an esoteric subject, and eventually the field became very crowded and competitive. It also meant the end of the antibody workshops. The Warner Springs meeting was to be the last of them in the United States. In 1966 a workshop was held in Israel, and by 1967 interest in immunology was so great that antibodies became the topic of the summer Cold Spring Harbor symposium and its big red book. The unspent funds remaining in the antibody workshops grant were returned to the NSF since that grant had obviously achieved its goal with a bang as the molecular biologists invaded the field. The Salk Institute, small and broke as it was, had played a key role in catalyzing the molecularization of immunology.

The three research areas selected for our first colloquia—neurobiology, the biology of aging, and molecular immunology—illustrate the pioneering intellectual role of the Salk Institute in the mid-1960s. These fields were hardly emerging, yet at the time they were not widely recognized as being accessible to inquiry and the wave of the future. Certainly they were not crowded, but soon they developed as main areas of research in biology, both at the Salk Institute and elsewhere.

The two weeks following our return to La Jolla after the Warner Springs Antibody Workshop were equally exciting. Two distinguished

FIGURE 8. Party at the Cohn home, February 1965. From left to right: Jacques Monod, Suzanne Bourgeois, Francis Crick, Leslie Orgel, Gobind Khorana. Author's private collection. Photograph by Mel Cohn.

visitors, Severo Ochoa and Gobind Khorana, joined the Fellows in stimulating seminars and discussion about the genetic code.[45] Moreover, following up on the inspiring February 1964 meeting on the nervous system, three members of the Harvard Medical School group had been invited to visit the institute and meet all the Fellows in February 1965. Stephen Kuffler, David Hubel, and his associate Torsten Wiesel presented and discussed their work for three days.

Under the leadership of Kuffler, that group had initiated the field of neurobiology by combining studies of brain anatomy, biochemistry, and physiology using techniques usually available only in separate departments. This interdisciplinary approach was entirely new at the time. They were attracted to our institute because they saw an opportunity to move their entire group and expand it while interacting with molecular biologists.[46] Hubel and his associates had been offered positions at Harvard College's Biology Department, but they did not want to split up the Harvard Medical School group.[47] Thanks to the extensive uncommitted and flexible space available in the Kahn building, joining our institute was a unique opportunity for them as well as for us. This was a lucky possibility that could have profoundly changed the development of the Salk Institute and of the field. The Fellows were unanimously enthusiastic, and in their February 24 meeting they agreed to communicate their opinion to the board of trustees, which was to meet a few days later.[48] Unfortunately, the timing could hardly have been worse.

The trustees' meeting woke us up to the reality of our disastrous financial situation and internal political upheaval. The financial crisis triggered by our dear Kahn building (see chapter 8) was turning into a chronic condition. In March 1963, when the National Foundation made funds available to cover operating expenses and institute deficits until 1966, that deadline had seemed so far away. By 1965, however, there was a desperate need for funds to outfit the north building, as the entire building loan had been spent and construction would have had to be stopped altogether by summer. With no funds in sight, it was clear that drastic measures had to be taken to save the institute. It was also clear, and ominous, that O'Connor and the National Foundation were getting impatient and frustrated with the inability of the Salk Institute's leadership to organize a constructive fund-raising program. It must be said that neither Jonas Salk nor Warren Weaver was experienced or gifted at raising money, but both showed superb judgment and taste in spending it. However, O'Connor made it clear that the National Foundation could not indefinitely support the deficit financing of the institute.[49]

Warren Weaver, as chairman of our board, proposed to bring into the institute as soon as possible a full-time and compensated president who would also be chief executive officer. This meant, of course, that Jonas Salk would have to resign as president. Moreover, Weaver introduced a person ready to accept that position, Augustus B. Kinzel, one of the institute's trustees. Kinzel had a B.S. in general engineering from MIT and a doctorate in metallurgy from the University of Nancy, France. He was one of the founding members of the National Academy of Engineering and its first president. He had a forty-year career at the Union Carbide Corporation, where he had risen to the position of vice-president for research. He had a home in La Jolla and would be ready to take up a position at the institute upon his retirement from Union Carbide on July 31, 1965. He would be in charge of the institute's administration and organize vigorous and vital fund raising. Jonas Salk, as director, would then be able to devote his time to the academic affairs of the institute and to his own research.

Kinzel began his presidency on August 1. He was the third institute president in a succession of twelve presidents to date.[50] We escaped bankruptcy as the National Foundation bailed us out once more. Fortunately,

FIGURE 9. The Founding Fellows, February 1966. From left to right: Francis Crick, Ed Lennox, Jacques Monod, Jonas Salk, Leslie Orgel, Mel Cohn, Salva Luria, Jacob Bronowski, Renato Dulbecco. Courtesy The Salk Institute.

research in the barracks was thriving, and Mel Cohn and I had travel plans to cheer us up: *revoir Paris!* This was our first return to Europe since our exile in La Jolla, and our timing was lucky. In October 1965, while we were spending time at the Pasteur Institute, Jacques Monod and his colleagues François Jacob and André Lwoff won the Nobel Prize in Medicine, and we were there to help them celebrate. By 1966 the labs were moved into the north Kahn building, as the entire institute staff now consisted of 153 employees; the barracks were leased to the UCSD Medical School and became a source of income. While Steve Kuffler accepted the chair of Harvard's new Department of Neurobiology, he also accepted a nonresident Fellow position at our institute.

The year 1967 started auspiciously with an important February colloquium, Studies on Development of Behavior.[51] It was meant to educate the trustees, faculty, and scientific staff so they might consider an eventual expansion of the institute into behavioral sciences. Two speakers did, in fact, have considerable impact on the institute's future: Daniel Lehrman and Eric Lenneberg. Lehrman, director of the Institute

FIGURE 10. Ursula Bellugi demonstrating sign language, 1974.
Courtesy The Salk Institute.

of Animal Behavior at Rutgers University, introduced us to the field of hormones and behavior, and he was to be elected nonresident Fellow in 1968 (see chapter 10). Lenneberg, a professor in the Department of Psychology at the University of Michigan, gave a talk titled "The Biology of Language Development," which discussed a field that was to become a major area of study at the institute.

While the Fellows were providing the institute president and the director with ideas and materials for the institute's academic plans,[52] it became clear that Kinzel did not interact well with the faculty and junior scientists. Admittedly, he did not have an easy job, and the late 1960s were turbulent times. The unpopular Vietnam War was escalating, and antiwar protest was spreading on campuses. The spirit of protest generated not only draft resistance but also the questioning of long-respected values and claims of rights to freedom of lifestyle and of expression. This contributed to the growth of the counterculture, including the use of drugs. It is in that social and political context that two unfortunate incidents rocked the Salk Institute and forced Kinzel's resignation. Those incidents were apparently unrelated except for their timing, merely days apart.

On April 18, 1967, a San Diego police officer accompanied by an agent from the Bureau of Drug Abuse Control of the U.S. Food and Drug Administration visited the institute to investigate a tip about the alleged manufacture of LSD in a Salk Institute lab. This came as a shock that was felt throughout the institute, and it was followed by uneasiness about how to react while the investigation was in progress.[53] Meanwhile, the director and Fellows had agreed that a local artist named Sandra Woodward-Baltimore would organize an art exhibit in an unoccupied space of the south Kahn building. Sandra was the wife of David Baltimore, then a vocal research associate in the Dulbecco lab and a prominent antiwar activist.[54] Due to open on May 1, twelve days after the visit by the police, the show included paintings and sculptures by three artists. The day before the opening, Kinzel previewed the exhibit and ordered that three pieces be removed because, in his opinion, they desecrated the American flag.[55] The artists threatened to remove all their works from the exhibit, and as a result the show was cancelled. The artists brought the incident to the press, and by the next day the story was in the newspapers and on television. The press commented on the Salk Institute's failure to maintain the courtesies and traditions of an academic institution, and this hurt the institute's reputation.[56] San Diego was a military town where recruits embarked for Vietnam while La Jolla was an academic community. The chairman of the Philosophy Department at UCSD sent an angry letter of protest to Kinzel.[57] Whether people sided with the artists or with Kinzel, everyone had a reason to be hostile, and this alienated the community. Didn't we have enough trouble?

The institute was torn apart from the inside as well, as most scientific personnel agreed with the artists, while others found the artworks objectionable. Moreover, the Fellows resented that Kinzel was able to overrule them in matters of academic policy. Eventually the issues were resolved, and the show opened in mid-May, but it was too late for Kinzel. On May 5, at a meeting of the executive committee of the board of trustees, he offered his resignation, which became effective in October. This ludicrous incident also tore at Jonas's heart. After the confrontation he wrote a memo intended for the entire staff, opening it with a quote from Adlai Stevenson, "I am too old to cry, and it hurts too much to laugh." He added, "I know now how he felt."[58]

CHAPTER 10

The McCloy Boys

Hope springs eternal.

—Alexander Pope[1]

In October 1967 the Salk Institute board nominated a new president. His name was Joseph (Joe) E. Slater, and he was to be CEO and trustee of the institute as well. Joe was tall, handsome, well dressed, a good talker in three languages (including German and French), and was extremely well connected. Within the first minute of meeting someone, he would mention enough names of powerful and famous people to make anybody incredulous. In fact, he did personally know an incredible number of celebrities in politics, business, and the arts. His connections were international, and he could perhaps be best described as a politico-cultural entrepreneur.[2]

Slater was born in Salk Lake City, Utah, in 1922 and grew up in Palo Alto, California, where he went to high school. After a stint in the U.S. Navy during World War II, he graduated from the University of California, Berkeley, with a B.A. in economics. After the war Slater became actively involved in the rebuilding and de-Nazification of defeated Germany. The Cold War made the world a complicated place, and Slater occupied a number of posts in Europe and in Washington, D.C. He worked for the Four Power Allied Control Council, the military occupation governing body based in Berlin, until 1948, when he moved to the staff of the State Department in Washington, D.C., to participate in setting up a division of the United Nations. In 1949, however, people who knew his work in Berlin helped send him back to Germany.

The military rule of West Germany ended in 1949, when a civilian high commissioner replaced General Lucius D. Clay, the U.S. military governor who had succeeded General Eisenhower and had so impressed Edward Litchfield (see chapter 2). John Jay McCloy was chosen for the job and sent to West Germany, then considered the center of the brewing Cold War.[3] McCloy was an extraordinary man who, even in his eighties, was described as the most influential private citizen in America. A graduate of Amherst College and Harvard Law School, he was a powerful corporate lawyer and presidential advisor. He was involved in some of the most tragic decisions the American president had to make during World War II. For reasons of national security, McCloy recommended the internment of the Japanese American community into concentration camps, but he later tried to convince Truman to further negotiate with Japan before dropping the atomic bomb. He had done war service in Washington, and by the end of World War II he was the assistant secretary in the War Department. He had acquired considerable experience and many contacts through his work at a Wall Street law firm and as president of the World Bank before being sent to Europe.

McCloy, however, did not know West Germany well, and he needed the help of a team of dependable advisors who were fluent in German and familiar with the region. His team included Slater, who had returned to Germany to join McCloy as secretary of the Allied High Commission. Another key member of McCloy's team in Germany was Shepard Stone. A distinguished journalist who had earned a doctoral degree in history in Berlin, Stone was on leave from the *New York Times,* and since 1945 had helped to rebuild the press in the U.S.–occupied zone. With the arrival of McCloy in 1949, Stone was put in charge of the large public affairs operation of the U.S. High Commission for Germany.[4] The trio—McCloy, Slater, and Stone—had much in common. All three were appalled by the destruction they found in Germany, and they wanted to help rebuild the country the right way—the American way. They all had an affinity for Germany—the good Germany of course, which Hitler had tried to destroy. Coincidentally, all three had German-born wives.

In 1952, McCloy's tenure as proconsul of Germany ended and the trio broke up. He took the position of chairman of the Chase Manhattan Bank while also working with the Ford Foundation on international affairs. Eventually McCloy became a trustee of the Ford Foundation, and by 1958 he was chairman of the board. He wanted Stone back in the United States and offered him the most important position of his career as director of the foundation's program in international affairs. Meanwhile, the Ford Foundation was becoming the richest philanthropic organization in the world. McCloy deserves much of the credit for securing an extraordinary financial legacy for the Ford Foundation by directing a highly successful Ford Motor Company stock flotation in 1956.[5]

Slater, on the other hand, moved to Paris in 1952 as executive secretary in the office of the U.S. representatives to the North Atlantic Treaty Organization (NATO) and the Organization for European Economic Cooperation (OEEC), which were established under the Marshall Plan. After living in Paris, he moved to Venezuela as chief economist for the Creole Petroleum Corporation, an American oil company. He also created that company's charitable foundation. In 1957, McCloy invited Slater to join him at the Ford Foundation in New York as the deputy director of its program in international affairs, which was headed by Stone, who became Slater's immediate superior.[6] This is where the trio were reunited as they took over the Ford Foundation, where Slater spent ten years. However, with the approval of McCloy, Slater took several lengthy leaves of absence from his job at the foundation. Both Slater and McCloy were convinced that, while working in the field of philanthropy, it was not desirable to immobilize oneself in one job. It was important to move in and out of philanthropy to broaden one's outlook and network, to get fresh ideas, and to develop new contacts.

Slater and Stone became known as McCloy Boys, which was the best recommendation they could have received.[7] McCloy Boys were a fraternity made up of people who formed a network of aggressive managers who had essentially been trained by McCloy in postwar Germany. They had superior experience and were sought after for important jobs because of the broad training and international background they had acquired while rebuilding a broken country. They were able people who, furthermore, had access to Ford Knox—pun probably intended—as the gold of the Ford Foundation was often called.

In 1967 the leadership of the Ford Foundation was turning over. McCloy had left his position as chairman of the board, and the foundation's president, Henry T. Heald, had been replaced by McGeorge Bundy. It was time for both Slater and Stone to consider other jobs as well. Slater had his first contact with the Aspen Institute for Humanistic Studies—and with the Salk Institute—in the summer of 1967. He needed some time away from the pressures of his New York office to work on a new project. He discussed this project with Alvin C. Eurich, an old friend he had known years earlier at the Ford Foundation. Eurich, then president of the Aspen Institute in Colorado, invited Slater to spend time in Aspen during the summer of 1967 as a scholar-in-residence to work on his project.

In 1950, Walter Paepcke, a Chicago philanthropist and industrialist who was chairman of the Container Corporation of America, founded the Aspen Institute for Humanistic Studies and became its first president.[8] Paepcke, a trustee of the University of Chicago, had been greatly inspired and helped by others affiliated with that university, including Robert Hutchins, the university's president.[9] In 1957, Paepcke created and filled the position of chairman of the Aspen Institute, leaving Robert O. Anderson to take over the presidency and develop its programs for more than fifteen years. Anderson was the chairman of Atlantic Richfield Company, America's largest rancher, and a philanthropist as well. Eurich took over the presidency of the Aspen Institute in 1963, when Anderson was elected chairman of the board. Unlike his two predecessors, Eurich was an educator who had held several positions in academia. He created the artists- and scholars-in-residence programs at the Aspen Institute in 1966.

When Paepcke first visited Aspen, in 1945, it was a decaying mining town. Paepcke, however, was seduced by the breathtaking scenery and saw the town as a dream setting for an American version of the Salzburg Musical Festival. He eventually helped finance an Aspen revival and founded his institute by attracting celebrities to a summer program of concerts, art exhibits, lectures, and seminars. By 1958, the Aspen Institute had expanded its program by inaugurating a winter executive seminar series for the discussion of timely issues. These seminar series soon drew business executives and government officials, making Aspen an attractive forum for leaders.

When Slater arrived in Aspen in the summer of 1967, his friend Eurich had resigned and been replaced by an interim president, William F. Stevenson. A friend of Hutchins, Stevenson had only agreed to fill the post until a permanent president could be found to replace Eurich. As a scholar-in-residence Slater attended an ongoing executive seminar, gave a lecture in the Aspen Institute Tuesday Night series, and was introduced to the Aspen Institute chairman Robert Anderson and other trustees. The lecture Slater gave was titled "Biology and Humanism."[10] By chance, Jerome Hardy, the publisher of *Life Magazine* and *Time,* attended Slater's lecture. Hardy, as a trustee and the president pro tem of the Salk Institute, was the chairman of the search committee to find a new president after Kinzel's resignation (see chapter 9). Hardy explored the possibility of Slater potentially accepting the presidency of the Salk Institute. Slater was interested and actively followed up.

In Slater's typical dynamic style, he rapidly produced for consideration by the Salk Institute's trustees a lengthy memorandum entitled "Essential Conditions to Ensure the Balanced Development and Vitality of the Salk Institute during the Next Four Years."[11] That memo mentioned the fusion of scientific, humanistic, and other disciplines to better understand the importance of the rapid changes in the life sciences in man's development and well-being. The Salk Institute board acted rapidly as well. By October the Salk trustees had expressed their full confidence in Slater and full support of the principles, procedures, operations, and general objectives outlined in his memorandum. Hardy, as Salk Institute president pro tem, had already prepared and signed a letter offering Slater a four-year appointment as president, CEO, and trustee to start on January 1, 1968.[12] That letter specified that the trustees considered Slater's direct stewardship of the nonscience activities and their fusion with the scientific activities essential to ensure the institute's vitality and growth.

Joseph Nee, trustee and CEO pro tem of the Salk Institute and a trustee of the National Foundation–March of Dimes (NF-MOD), then reported the good news. At a meeting held that same January morning, the NF-MOD had approved a generous incremental grant to the Salk Institute. That generosity came with a few conditions, one of which was that Slater serve as president during the four-year period of the grant. Moreover, the Gildred Foundation immediately authorized another special grant to complement the NF-MOD grant. Suddenly the Salk Institute was rich. The deed was done—almost.

The Slaters arrived in La Jolla in January 1968. A simple note typed on Salk Institute stationery invited the senior staff—with their spouses—to meet Mr. and Mrs. Joseph E. Slater at a reception to be held at the institute on January 23.[13] It is on that invitation that Jonas Salk wrote by hand:

Mel
    Hope springs eternal!
    Jonas

All the nonresident Fellows arrived immediately after Slater since the regular annual Fellows meeting was scheduled for February 11–12. The Fellows invited the new president to attend their annual meeting. They had been discussing an academic plan since February 1967 and were prepared to present Slater with their wish list.[14] However, that was not quite the way Slater operated. He announced that a planning exercise would start on February 15 to carry out an intensive review of the current status of the institute and of its future.[15] He outlined the areas to be studied by a number of working groups. That planning exercise would last at least two months, resulting in recommendations that would be presented to the board in May in the form of a program of action.

As expected, Slater introduced new administrative staff members. A candidate for the very first Salk Institute vice-president position arrived on February 4 and met with all the Fellows and a few junior scientists. He seemed perfectly suitable, even appealing. Born in Oklahoma, John C. Hunt had graduated from Harvard College and continued his studies at the University of Iowa and the Sorbonne. He was fluent in French and had traveled extensively. He had taught English, Greek, and American history and was a novelist who had already published two award-winning novels.[16] Since 1956 he had served as the executive secretary for the Congress For Cultural Freedom (CCF) in Paris, and by the time he resigned from that organization in December 1967 he was the CEO.

The CCF was a liberal organization that had been established in Berlin in 1950 by distinguished European and American intellectuals to counter oppression by all totalitarian regimes. They opposed both Hitler and Stalin. Post–World War II Berlin was squeezed between defeated Nazism and blossoming communism; West Berlin appeared as an enclave of freedom. When the Soviets imposed the West Berlin blockade, Berliners were saved from starvation by the American airlift ordered by General Lucius Clay (see chapter 2). The airlift lasted until the fall of 1949. West Berlin in 1950 was the natural place to inspire an organization such as the CCF. Eventually the CCF created regional offices around the world, with headquarters in Paris. It sponsored conferences, festivals for artists and writers, and the publishing of prestigious intellectual magazines, including *Encounter*.[17] Michael Josselson—who in 1949 had briefly joined the McCloy-Slater-Stone trio as a cultural officer at the U.S. High Commission in Berlin—was the administrative secretary of the CCF in Paris and the direct superior of John Hunt.

Soon it became known that some of the younger Salk Institute scientists were uncomfortable about hiring John Hunt. The reason was that the CCF had been in the middle of a scandal in April 1966 when the

FIGURE 11. Reception at the Hunt's residence on the occasion of the annual Fellows meeting, February 1970. From left to right: Joseph Slater, Ann Slater, Salva Luria, Chantal Hunt, John Hunt. *La Jolla Light.*

New York Times had revealed that the CCF had received covert CIA support channeled through various foundations, including the Ford Foundation.[18] Although this was no longer the case, that information did lead to the resignation of Michael Josselson in 1967 and the demise of the CCF. It was, however, replaced by a similar organization under a slightly different name—the International Association for Cultural Freedom (IACF), which was headed by Shepard Stone and had no ties to the CIA. Stone wanted to keep John Hunt as his top associate after the scandal, but Hunt felt uncomfortable in that position and resigned by the end of 1967. This was the same time Joe Slater was considering the job as president of the Salk Institute.

John Hunt was not a McCloy Boy since he was not in postwar Berlin, and he knew Slater only slightly, through his contacts with the Ford Foundation while at the CCF in Paris. However, the fact that Shepard Stone wanted Hunt to remain at the IACF was, as far as Slater was concerned, the strongest recommendation he could have. Aware of what had happened to the CCF and that Hunt was looking for a job, Slater wanted him to be considered for the vice presidency at the Salk Institute. While Hunt favorably impressed the Fellows, the junior scientists were grumbling, which was worrisome. It appeared that the antiwar activists assumed that Hunt must have been a supporter of the official American position on the Vietnam War since he had worked for an organization financed by the CIA. This had to be dealt with quickly for fear of creating another incident similar to the art show episode that had toppled Kinzel one year earlier (see chapter 9).

It was Monod, a born diplomat, who suggested an approach. He proposed to invite some of the most vocal junior scientists to an informal gathering away from the Salk Institute. A casual dinner was arranged at the Cohn residence for the usual protestors, including Martin, Chuck, Dave, and Paul, to mention a few.[19] Jacques Monod, Francis Crick, and Mel Cohn represented the Fellows, and Hunt was asked to make a short presentation on his career and his outlook on Vietnam. At that point he was grilled by the junior scientists and this cleared the air. Hunt and his charming wife, Chantal, conquered everybody, and on the next day Hunt became the first Salk Institute vice-president.[20] His title was "vice-president for operations and new activities," but he was soon to become the institute's first executive vice-president.

However, not all members of Slater's administrative team fit in as well as John Hunt. A number of characters appeared who did not seem to belong at the institute. They were dressed in business suits and did not mix with the rest of the staff. The Salk scientists were soon to find out that they were ghostwriters. Scientists in the 1960s were quite naïve. They knew, of course, that politicians used speechwriters and ghostwriters, but scientists took pride and joy in writing their own papers, talks, books, and grant applications. Salva Luria's reaction to an incident illustrates the opinion of scientists about ghostwriting at the time. In the late 1960s, Luria was in charge of organizing at the Salk Institute a series of workshops on the mode of action of drugs and the problems they create. He sent a furious letter to John Hunt when he found out that a grant application to finance those workshops had been written without consulting him and was ready for his signature.[21]

The final program of action submitted to the board in May 1968 was a spectacular example of the kind of scientific ghostwriting used at the time.[22] More than writing, it was the attractive packaging of information provided by the Fellows, the president and his staff, and expert consultants, including architects. The final product was a set of four colorfully bound booklets—blue, yellow, pink, and green—entitled "I. Programs," "II. Organization and Operations," "III. Facilities and

Finances," and "IV. Fund-Raising." Not to mention volumes of supporting information.

Some forty-five years later that program is still impressive. Ambitious and idealistic, it proposed that the Salk Institute could become a very special place that would have a much broader impact than a purely scientific research institution. It suggested explorations into new nonscientific fields, reaching toward the behavioral and social sciences and the humanities. Initially these new programs were to be the direct responsibility of the president.[23] The program proposed probing into the social and humanistic implications of biological discoveries and studying, for example, how the biological revolution affected human thought and human values. Specifically, the program would explore the biological, psychological, and social nature of drug addiction, an area of special interest to Luria. In some ways this program's concern for social and political affairs would also have pleased Leo Szilard. It proposed the examination of whether cooperation in the field of biology could serve international relations and foster public awareness about biological warfare in particular. The eight-year program of action also considered building additional facilities, including a meetinghouse and visitors' residences as originally planned by Kahn (see chapter 8). All the program needed was a president able to raise the necessary funds and entirely devoted to the success of the program. Unfortunately, that was not to be Slater—or anyone else.

The first year of Slater's presidency was promising. By May 1968 the board approved the program of action and granted the Fellows several of their most pressing wishes. The board confirmed the appointment of Danny Lehrman as nonresident Fellow,[24] while Robert Holley was appointed resident Fellow. A sign of the Salk Institute's growing reputation and good judgment was the fact that in less than two months after his appointment, Bob Holley received the Nobel Prize.[25] The board approved another crucial step in the development of the Salk Institute, namely the creation of junior faculty ranks that were designated as "Members" and "Junior Members." Junior faculty would soon share with the Fellows the responsibility for academic decisions.[26] Finally, in December, John McCloy was elected trustee, and he would soon become chairman of the board to fill the position that had essentially been left vacant since 1965 by the ailing Warren Weaver. The year 1968 definitely ended on a high note.

As far as funding was concerned, Slater had close contacts with private foundations but relatively limited access to federal funding

FIGURE 12. Fellows and President Joe Slater, 1969. From left to right: Jonas Salk, Francis Crick, Joe Slater, Bob Holley, Warren Weaver, Steve Kuffler, Jacques Monod, Mel Cohn, Ed Lennox, Salva Luria, Danny Lehrman, Jacob Bronowski, Renato Dulbecco, Leslie Orgel. Courtesy The Salk Institute.

agencies. The Fellows were very successful at raising research money from the National Institutes of Health, the National Science Foundation, the U.S. Army, and even NASA. However, to ensure expansion, more federal resources had to be tapped. After all, they had access to the real Fort Knox! Slater quickly figured out an approach. To attract additional federal support, he simply hired a consultant who had contacts in government and knowledge of Washington trends and politics related to the support of scientific research. That consultant set up a Salk Institute office in the U.S. capital, where his job was to monitor the Washington scene, report to Slater on possible sources of funding, and introduce him to key people.[27] One productive contact made in Washington was Dr. Philip R. Lee, then assistant secretary for health and scientific affairs at the Department of Health, Education, and Welfare (DHEW). At the time, Dr. Lee was a physician who was the government's top biomedical official.

Dr. Lee was personally interested in the problem of population control and fertility. Reproductive physiology was a research subject that attracted support not only from large private foundations but also from government sources, in particular the U.S. Agency for International

Development (USAID). Birth control had also been a major area of investigation proposed for the institute by Leo Szilard (see chapter 4), although he was unable to obtain financial support to found an institute on that basis. None of the founding Fellows of the Salk Institute were working in that area, as they were more interested in developing the new field of neurobiology, but they were willing to learn about the state of the art in reproductive physiology. The challenge was to identify programs that would be compatible with the basic research going on at the institute and still sufficiently related to the population problem to attract funding.

A workshop on the subject was held at the Salk Institute on January 14, 1969, to which all scientific staff were invited.[28] The guest speakers were three leading scientists who had been invited as consultants to bring the fellows up to date on the field: C. H. Li, director of hormone research, University of California, San Francisco; Joseph Goldzieher, director of the Southwest Foundation for Research and Education, San Antonio; Roger Guillemin, director of the Laboratories for Neuroendocrinology, Department of Physiology, Baylor University, Houston. This was a remarkable meeting that lasted all day, changed the Salk Institute, and probably saved it from extinction. All three speakers were distinguished scientists, but it was Roger Guillemin who provided the link between the different areas into which the institute wanted to expand, namely reproductive physiology, neurobiology, and the study of behavior.

Born in Dijon, the capital of Burgundy, France, Guillemin was proud to be a *bourguignon*.[29] He completed five years of medical study in Dijon, two of which were during the difficult and sad period of Nazi occupation. The medical curriculum in Dijon was entirely clinically oriented, and Guillemin practiced medicine for a while after his studies, although he was already interested in endocrinology and hoped that he might work someday in a laboratory. When he heard that the pioneering endocrinologist Hans Selye was lecturing in Paris, he went to hear him and meet him. A few months later Guillemin went to Montreal, where he worked for a year in Selye's new Institute of Experimental Medicine and Surgery at the university. He completed enough experimental work to satisfy the requirements to obtain an M.D. degree from the Faculté de Médecine of Lyon in 1949. With his degree, Guillemin promptly returned to Selye's institute, where he learned experimental endocrinology. After obtaining a Ph.D. degree in physiology from the University of Montreal in 1953, he then joined the faculty of the Department of Physiology at the Baylor University College of Medicine, where he taught physiology for eighteen years while developing a very active endocrinology research group.

By the time Guillemin first visited the Salk Institute in January 1969, his group at Baylor had been working for several years on brain factors that act on the pituitary gland to release hormones that regulate the activity of the ovaries and testes. The hypothalamic area of the brain was involved in the production of those hormone release factors. Guillemin created a sensation during his seminar when he casually mentioned that fifty tons of sheep hypothalamus fragments collected in slaughterhouses had been handled in his laboratory to purify minute amounts of one of those factors, TRF (thyrotropin-releasing factor). That evening over dinner and a glass of wine, Salk scientists invented a new unit, the *guillemin,* corresponding to one ton of sheep hypothalami![30] Danny Lehrman, the most recently appointed nonresident fellow and a pioneer on the study of mating behavior in birds (see chapter 9), was put in charge of reviewing the accomplishments of Guillemin and other candidates.

At the Fellows meeting in February of 1969, Lehrman reported that Guillemin was not only a first-rate scientist but he could also relate well to the Fellows.[31] His studies of hormone release factors would create a bridge between molecular biology and behavior. He pointed out that since the institute wanted to expand into neurobiology and reproductive biology, research on the hypothalamus was at the point where those two fields merged. Mel Cohn, who as chairman of the faculty acted as chairman of the Fellows meeting, reported on the January workshop and on a proposal he wrote on the use of tumors of the hypothalamus as a possible enriched source of release factors. Cohn suggested that Guillemin might be invited to visit while all the Fellows were in La Jolla. Jonas Salk was put in charge of contacting Guillemin, who happened to be in California at the time. On February 10 Guillemin visited the institute and met all the Fellows, a visit he remembered fondly.[32] He was moved by the beautiful Kahn building, impressed by the large empty space in which he could build a lab of his own design, and struck by the gracious president Slater, who seemed able and eager to raise the necessary funds. In addition there was an impressive group of Fellows and administrators, most of them who spoke French with a variety of accents, including Monod, Luria, Lennox, Cohn, Slater, and Hunt. Guillemin simply could not resist. He was ready to move and to convince his entire group of junior collaborators to move with him.

By October 1969 the Salk Institute Board had approved the appointment of Guillemin and gave to the administration the authority to proceed with negotiations to appoint him as a Fellow, effective July 1970.[33] Funds were assured, the space had been agreed upon, and Guillemin had convinced key members of his group to join him in La Jolla to found the new Neuroendocrinology Laboratory at the Salk Institute. Morale was high as a major expansion of the institute was planned, involving the activation of new lab space on the first floor of the south Kahn building. Moreover, following up on the creation of junior faculty ranks, the first five junior members were appointed by the board in February. They were Ursula Bellugi, Suzanne Bourgeois, Walter Eckhart, Stephen Heinemann, and David Schubert. Bellugi founded the Laboratory for Language Studies (see chapter 11), while I founded the Regulatory Biology Laboratory and Eckhart took over the Tumor Virus Laboratory after the departure of Dulbecco. Heinemann and Schubert founded the first Neurobiology Laboratory at the Salk Institute.

After the failure to recruit the Harvard Neurobiology Group (see chapter 9), the Fellows conceived a more modest plan that involved attracting promising young scientists from other fields and training them in the neurosciences with the help of the Harvard group. Stephen Kuffler was appointed nonresident Fellow, and funds were obtained to host Kuffler and his colleagues at the institute for several summers to educate the faculty and junior scientists in the new field of neurobiology. Dave Schubert was the first UCSD graduate student to receive a Ph.D. for work done at the Salk Institute.[34] After he completed his work in the Cohn lab, he spent a postdoctoral period with François Jacob at the Pasteur Institute in Paris. Steve Heinemann had obtained his Ph.D. at Harvard with Matt Meselson and completed postdoctoral work at MIT and Stanford. Schubert and Heinemann had taken summer courses taught by the Harvard group, and they established the first neurobiology lab at the Salk with additional help from Kuffler. Three junior neurobiologists recommended by Kuffler eventually joined them: John Harris, Yoshi Kidokoro, and John Henry Steinbach. Thanks to Stephen Kuffler, the new neurobiology program at the Salk Institute quickly reached a critical mass and was productive starting in the early 1970s. After the setback of being unable to recruit the Harvard group, a neurobiology program at the Salk was back on track. However, a major crisis was afoot.

The first signs of trouble had already appeared in February 1969, at which time nobody seemed very concerned.[35] At the board meeting Slater had announced his intention to move his headquarters to New York. This was reasonable since the East Coast is generally the financial center of the country, and Slater would have to spend much of his time raising funds. He proposed that John Hunt, as executive vice-president residing in La Jolla, would administer the institute, having full authority over its day-to-day management. This was agreeable to everyone, and by June Slater had already sold his residence in La Jolla and moved back to New York. More worrisome was Slater's proposed arrangement that he also preside over the Aspen Institute.

It became obvious by February 1969 that Slater had already carried out extensive discussions with members of the Aspen Institute board, in particular with Robert Anderson, its chairman. In addition to the presidency of the Aspen Institute, Anderson had proposed that Slater assume responsibility for creating and directing a new foundation, the Anderson Foundation. Slater viewed all of these roles as not only compatible with but complementary to his Salk Institute function as president. He envisioned possible ties between the Aspen Institute for Humanistic Studies and the Salk Institute for Biological Studies, and he would share his time between the two institutes. At the time this seemed appropriate and even attractive, as novel ideas and new funding for programs bridging the sciences and the humanities could be generated from interactions between the two institutions. However, by October 1969 Slater had informed the Salk Institute board that the time had come to find a successor, as he wanted to assume the presidency of the Aspen Institute by March 1, 1970.[36]

By February of 1970 the crisis had come to a head. The National Foundation suspended payment of its grants, placing the institute in what Chairman McCloy aptly described as a "state of suspended animation."[37] Lawyers got into the act and in May informed McCloy that all NF-MOD grants were cancelled since they were conditional on Slater remaining president of the Salk Institute for the four years of the grants. There was enormous pressure on Slater to remain for the full four years, and, to make matters worse, John Hunt was due to leave in July. This is when Slater proposed to stay as half-time president with the help of a chancellor. He had a candidate in mind: Frederic de Hoffmann (see chapter 5).

De Hoffmann was elected chancellor against the will of the National Foundation. The four trustees representing the NF-MOD who were sitting on the board of the Salk Institute abstained from voting since they considered that no president, no matter how able, could do the job working only part-time. The other Salk Institute board members yielded to the practical consideration that Slater's resignation could endanger millions in grant applications that were in the pipeline and were predicated upon Slater being president. Moreover, even if Slater remained as part-time president, there was still a chance that the NF-MOD might reverse its decision and pay the incremental grants. That was never going to happen, however, as long as O'Connor was at the helm of the NF-MOD. What happened was a tragic breakdown in the relationship between the NF and the Salk Institute in the midst of what should have been happy days. Indeed, while difficult negotiations between the NF-MOD and Salk trustees were taking place, Jonas, who had long been separated from his first wife, Donna, married the talented, beautiful, rich, and famous French painter Françoise Gilot.[38]

Jonas and Françoise first met in October 1969 in La Jolla through John and Chantal Hunt. Françoise and Chantal had known each other in Paris in the mid-1950s.[39] At the time, they each had two children, a boy and a girl, and the four attended the same school in Paris, the Ecole Alsacienne.[40] The two mothers became good friends and had stayed in touch over the years. In the spring of 1969, while John and Chantal Hunt were living in La Jolla, Françoise happened to have a show in Los Angeles and contacted Chantal. Françoise returned to Los Angeles in September of that year and asked Chantal to visit so they could spend a few days together, which they did. In October, Françoise came down to La Jolla and stayed with the Hunts for a long weekend. On Monday, John brought Jonas Salk home for lunch, and on Tuesday they met again at a Salk Institute dinner held at a local restaurant. Françoise and Jonas seemed to have little in common until the next day, when Jonas called to pick her up to visit the Salk Institute. Françoise was captivated by the Kahn building, and suddenly the two had something beautiful and interesting to talk about.

At that time Françoise was spending about three months of each year in the United States, and she spent another month in New York after her stay in California. To her astonishment, Jonas showed up in New York in November and wanted to visit her in Paris in December to "see the trees illuminated on the Champs-Elysées for Christmas!" Jonas did, however, have a somewhat better excuse for visiting: he wanted to see Jacques Monod. Knowing that Monod had a house in the hills above Cannes where he would most likely spend the holidays (see chapter 6), Françoise invited Jonas to her house in Vallauris, a little town less than two miles from Cannes, where they spent the holidays together and visited Monod. They met once more in the United States in early 1970, when Jonas told her that he intended to marry her. She quickly found out what the founders of the Salk Institute knew so well—that Jonas was persistent. When she went on a cruise to Greece Jonas called her on the yacht in the middle of the Aegean Sea. Back in New York, they realized that their relationship had made the papers and they were assaulted by paparazzi. To avoid the press they were married in France on June 29, 1970, in Neuilly-sur-Seine. John and Chantal were their witnesses. The newspapers announcing their wedding did not miss the chance to mention that some twenty years earlier Madame Gilot had been the longtime companion of Pablo Picasso, with whom she had two children out of wedlock. This, however, was no secret since Françoise herself had written an autobiographical best seller covering that period of her life.[41]

They returned to the United States in the summer, Jonas stopping in New York to see O'Connor and Françoise going on to Los Angeles, where she stayed with friends until Jonas joined her so that they would arrive in La Jolla together, as newlyweds should.[42] No one will ever know what was said between Jonas and O'Connor that summer of 1970, but it appears that this meeting marked the end of their relationship. This was precisely the time when de Hoffmann was assuming his new functions as chancellor. By then, all the NF-MOD grant payments had stopped and the cash available to the Salk Institute was limited to one payroll. Meanwhile, more than a half a million dollars in unpaid bills had accumulated.[43]

Fortunately, Delbert (Del) Glanz was appointed treasurer as it became clear that the institute was entirely broke. The institute survived on loans, letters of credit, pledges, austerity, and probably Del's creative accounting, which required borrowing from Peter to pay Paul. Meanwhile, Roger Guillemin had arrived at the Salk Institute in June only to find out that Slater was essentially gone and Chancellor de Hoffmann had to be dealt with. The construction and move into Roger's beautiful new laboratory in 1971 turned out to be more difficult and less pleasant than anticipated.[44]

The negotiations between the NF-MOD and the Salk Institute had reached a stalemate, and by September 1971 O'Connor and Nee had left the institute's board. At the board meeting of February 1972, as planned, Slater officially resigned four years after his appointment as president. It was at that stormy meeting that de Hoffmann was elected to succeed him as president. He turned out to be the only contender since it was obvious that in its disastrous financial situation, the institute would not attract sought-after candidates. Moreover, de Hoffmann, who had lost his position at General Atomic, needed a job as much as the Salk Institute needed a president (see chapter 12).

On March 9, four weeks after Slater's resignation, O'Connor died while in Phoenix to attend an NF-MOD Science Advisory Committee meeting.[45] Sadly, it appears that this was the only way the negotiations between the NF-MOD and the Salk Institute could resume. By May, a meeting had taken place between representatives of the NF-MOD and the Salk Institute. The atmosphere was described as cordial, and the meeting resulted in a formal letter proposing a special grant to the institute with very reasonable conditions.[46] The crisis was over, but there was still more bad news to come in 1972. That spring Dulbecco negotiated an extended leave of absence and left for London to join the laboratory of the Imperial Cancer Research Fund. Regardless of the reason for his departure, it certainly looked like he was abandoning ship.[47] Then, in August, Danny Lehrman died of a heart attack at age fifty-three. The Salk Institute was in a state of shock and was never quite the same after 1972.

The Slater affair remains baffling to this day. Why did Slater abandon the Salk Institute presidency after only one year even though he appeared so successful in the role?[48] Most likely, he came to realize that his future was not in academia. He aspired to become an international political leader, and the presidency of the Salk Institute would not help him achieve his personal ambitions. He wanted to be connected to government and business power. The Aspen Institute could help him reach this goal, and he did not want to miss the job opportunity offered by Anderson.[49]

Given Slater's presumed agenda, one might wonder why he accepted the Salk Institute presidency in the first place. In some ways, however, leadership of the institute was an attractive opportunity for an administrator. The institute had several important assets, including a remarkable building that was largely empty, a location in a growing intellectual center, a recognizable name, the ability to attract first-rate minds doing cutting-edge biological research, and a commitment from the NF-MOD. However, the institute, which was still in the early stages of development, needed consolidation, and a creative administrator would be able to leave his mark.

However, a pure research organization like the Salk Institute needed an endowment to survive because it received no income from tuition, alumni, or a clinical facility. Slater had been successful at raising money for specific programs but not at obtaining money for the institute's endowment. According to John Hunt, this is what made O'Connor so angry that he risked destroying the institute in which he had already invested so much effort and NF-MOD money.[50] To make it clear that the support of the NF-MOD was not equivalent to an endowment and that he was in control, O'Connor simply cut off payment of the grants. Moreover, O'Connor had reasons to feel that the NF-MOD had been tricked into offering generous incremental grants based on Slater's commitment to a four-year presidency. The rupture between O'Connor and Salk in the summer of 1970—after almost twenty years of sharing affection and dreams—could have had other reasons as well. Perhaps Jonas defended Slater because he viewed him as an asset that would help the institute create a bridge between science and the humanities. Moreover, Jonas's marriage to Françoise did not sit well with many because foundations that depend on fund raising feel threatened by any hint of scandal that could put off donors.

One can only imagine what might have happened if Slater had remained the Salk Institute president for four years or longer. The Salk Institute would be a different place today, perhaps with broader goals and a deeper impact than a pure research institute. As things happened, Slater left the Salk Institute with a mixed legacy. Making possible Guillemin's move into the south building was the most positive result of his presidency, and the creation of the Council for Biology in Human Affairs also had temporary value (see chapter 11). However, the takeover by de Hoffmann killed the original spirit of Jonas's institute (see chapter 12) without providing a solid financial basis for its expansion.

Biology in Human Affairs

Homo sum: humani nihil a me alienum puto.

—Terence[1]

From its inception, the Salk Institute was conceived as a center for research sensitive to the social and humanistic implications of advances in biology. Leo Szilard was a symbol of the scientists' social responsibilities. The destructive use of atomic energy in World War II made him painfully aware that scientists in general—not just physicists—were accountable for their discoveries. Biologists in particular research problems that raise a variety of issues that affect the well-being of man. This makes biology the ideal discipline for building bridges between the two cultures, that of the sciences and that of the humanities. That was the reason Jonas Salk had contacted Jacob Bronowski (Bruno) in 1960 and invited him to join the institute as a founding resident Fellow in 1962 (see chapter 7).

Bronowski's interest was human specificity, or the attributes that make humans unique among animals.[2] He was particularly interested in human language and since 1964 had been in touch with Roman Jakobson, the distinguished linguist, about the possibility of becoming a visiting Fellow of the Salk Institute. Born in Russia and capable of speaking numerous languages, Jakobson had a great impact on the field of linguistics, holding joint appointments at Harvard and MIT. He had been drawn to biology in the 1960s when the genetic code had been deciphered. He was interested in having closer contacts with molecular biologists, and so, during the summer of 1966, after the laboratories had been moved from the barracks into the north Kahn building, Jakobson spent two months at the Salk Institute.[3]

Jakobson gave a number of exciting seminars, and, as a result, the resident Fellows invited the linguist Eric Lenneberg to give a lecture entitled "The Biology of Language Development" at their next colloquium in February 1967, when the nonresident Fellows would also be in La Jolla (see chapter 9). Lenneberg, a professor in the Department of Psychology at the University of Michigan, had pioneered ideas on language acquisition. Jakobson spent another two months at the institute in 1968, and he participated in the meeting The Biological Foundations for Language, organized by Bruno in the spring of 1969.[4] Those early contacts with linguists introduced Ursula Bellugi to the institute. Ursula, one of the first junior faculty members, appointed in 1970, was concerned with the acquisition of language by children (see chapter 10). As her studies became integrated into some of the neurobiology programs of the Salk Institute, the Laboratory for Language Studies that she founded eventually evolved into today's Laboratory for Cognitive Neuroscience.

As the Salk Institute's in-house humanist, Bruno contributed much to the intellectual life of the institute. In addition to attracting remarkable linguists, he invited the renowned philosopher of science Karl R. Popper, who spent two months at the institute as a visiting Fellow in the winter of 1966. Popper struck up friendships with Jacques Monod and Mel Cohn. Monod ended up writing the preface of the French translation of Popper's famous book *Logic of the Scientific Discovery*,[5] while Cohn used his conversations with Popper as the basis for a lecture to the Society for Scientific Temper in India.[6]

One of Bruno's early tasks was to organize the institute library. He established not only a standard working library for the institute's scientists but also a very special collection of books reflecting the history of our understanding of the brain and the central nervous system, initiated in memory of Leo Szilard after his death in May 1964 (see chapter 9). Far from being a museum of rare books, this was to be a unique working library on the history of the neurosciences from the 1500s to the present. The Szilard Memorial Collection, purchased with funds collected by the Women's Association for the Salk Institute (see chapter 8), included a remarkable set of rare books, some dating back to the sixteenth century.[7]

The Szilard Collection was to be part of a broader Archive of Contemporary Biology, which was to serve both educational and historical purposes.[8] The archives were to contain reprints, films of lectures and recordings of interviews by resident and nonresident Fellows and other

FIGURE 13. Karl Popper at the Salk Institute, 1967. Courtesy The Salk Institute.

leading scientists, and historical material on the Salk Institute. In addition, there was instituted a series of occasional papers, which were distributed gratis by the institute and intended to capture the attention of a select audience.[9] Unfortunately, most of these humanistic projects were aborted after Bruno's untimely death in 1974 (see chapter 12). However, the major role he played as director of the Council for Biology in Human Affairs (CBHA) left a durable legacy that is mostly forgotten today but profoundly changed the lives of a number of people and left a lasting social legacy.

While he was president of the Salk Institute, Joe Slater created the CBHA after the board approved the program of action (see chapter 10). The CBHA was to be the nexus of a network of individuals and organizations sharing their concerns about the consequences of the rapid progress being made in biology. Formally established in 1970, the council counted international leaders from business, finance, and politics as well as distinguished scientists and humanists among its members. Bronowski, Holley, Luria, Monod, Salk, and Slater represented the Salk Institute. Operating under the belief that forming public opinion is best done by informing it, the council had as one of its major goals identifying problems and communicating its ideas and results to public leaders.[10]

The CBHA consisted of six commissions, each chaired by a member of the council:

Biology in International Affairs (Paul Doty chair)

Biology, Ethics, and the Law (Abram Chayes and Joseph Goldstein cochairs)

Biology, Learning, and Behavior (Eugene Galanter chair)

Biology in Contemporary Culture (Jacob Bronowski chair)

Ecology, Environment, and Population (Cyrus Levinthal chair)

Biology, Medicine, and Health Care (chair to be selected)

The council met regularly, either in La Jolla or New York, to identify problems, attend seminars, discuss options, and plan communications to the public and its leadership regarding new biological knowledge and its implications for human affairs.

Bronowski accepted the position of director of the CBHA from 1970 to 1971, despite the fact that he had just started to write the outline of *The Ascent of Man,* the BBC television series and book that were to be his magnum opus. Harry Boardman was appointed secretary general.[11] For several years Boardman had been in charge of organizing meetings at the Council on Foreign Relations, an influential think tank that specializes in U.S. foreign policy and international affairs.[12] He had extensive experience organizing lectures, symposia, and seminars, as well as small meetings that fostered private conversations between council members and statesmen, political leaders, and scholars. He was to do a similar job for the CBHA, which was clearly modeled on the executive seminars offered at the Aspen Institute (see chapter 10). Bronowski, meanwhile, ensured the quality of the programs supported by the CBHA. Slater had been able to raise funds to finance a variety of council activities.[13] One of the interesting uses of those funds was to support unusual postdoctoral fellowships.

An interesting example of such a fellowship was that of Michael (Mike) F. Jacobson. After graduating from the University of Chicago in 1965, Jacobson moved to San Diego on the advice of George Beadle. He joined the brand new UCSD graduate school, which allowed students to spend research-training periods at the Salk Institute.[14] Jacobson elected to work with David Baltimore in the barracks, becoming, along with

Dave Schubert, one of the first two UCSD grad students at the Salk. Jacobson followed Baltimore to MIT in January 1968, where he completed his Ph.D. in 1969. He had done outstanding work on the formation of poliovirus proteins in Baltimore's lab, and although he could have obtained a position in any lab he chose, Jacobson was not sure that was what he wanted.

The late 1960s were a time of civil rights debates and anti–Vietnam War demonstrations. Those activities prompted Jacobson to consider taking a year away from academia to use his scientific background to solve social problems in a way that was more direct than working on some esoteric basic research question. He was attracted to what the activist Ralph Nader was doing, and he wanted to get involved to help. Eventually Jacobson managed to meet Nader, who accepted him as a volunteer at his Center for Study of Responsive Law. As this was an unpaid position, Jacobson badly needed a postdoctoral fellowship. However, even with a B.S. in chemistry from the University of Chicago and a Ph.D. in microbiology from MIT, obtaining a fellowship to work outside a lab was essentially impossible. Somehow, through Baltimore's contacts, Jacobson heard that Cyrus Levinthal, a biology professor who had moved from MIT to Columbia University in 1968, had access to unusual fellowship funds from the Salk Institute. Levinthal, as chairman of one of the CBHA commissions, provided Jacobson with a fellowship that allowed him to work with Nader for a year.[15]

Since he left the Salk before its creation, Jacobson had actually never heard of the CBHA. He only knew that his modest paychecks came from the institute. That fellowship was invaluable to him, allowing him to move to Washington and explore working on social issues outside the lab. Those events actually changed the course of his life. In Nader's group his job was to research and write about food additives. This led him to write a book—*The Eater's Digest*—and encouraged his interest in nutrition.[16] In 1971 Jacobson and two other scientists he met in Nader's group set up their own organization, the Center for Science in the Public Interest (CSPI).[17] Since then Jacobson has worked very successfully to improve government policies and corporate practices related to diet and health. Today the CSPI, with Jacobson as its executive director, is the most active and influential proconsumer food lobby in Washington. Among their many achievements, Jacobson and his organization led the efforts to get nutrition labels on foods, calorie labeling on chain-restaurant menus, stronger food-safety laws, and artificial trans fat out of most foods. Mike is personally credited for coining the term "junk food."

Another unusual postdoctoral Fellow who was involved in the activities of the CBHA was Michael Crichton. In 1969 he had just earned his medical degree from Harvard and had started working in Jonas Salk's lab. Three years earlier he had entered medical school and started writing articles and novels under pseudonyms to pay for his tuition. His first best-selling novel, *The Andromeda Strain,* was published the year he arrived at the Salk Institute.[18] Published under his own name, the book made him an overnight star as an author of science-fiction medical thrillers.

As both a medical doctor and a writer, Crichton successfully built a bridge between the two cultures, reflecting the fact that biologists and humanists ask similar questions about man. Well-researched science fiction engages a wide audience and inspires curiosity and excitement about the possibilities offered by scientific research and technologies. After his year at the Salk, Crichton became a full-time writer, primarily of science fiction novels, many of which have been made into movies or television shows. However, he also wrote nonfiction books. In fact, while at the Salk he gave several talks about medicine in our Speaking-on-Tuesday seminar series.

Director of the CBHA and chair of the Commission on Biology in Contemporary Culture, Bronowski invited Crichton to participate in two symposia on The Entry of Biology in Contemporary Culture. At the second symposium, held in New York City, Crichton presented a lecture titled "A Contemporary Account of Human Nature."[19] Though Crichton's participation in Salk Institute activities is generally unknown, Crichton is remembered fondly by some of the old-timers at the Salk Institute.

The CBHA considered it particularly important that young scientists at the Salk Institute form working groups to discuss the social impacts in their own areas of research, and that these groups be treated as active parts of the council's six commissions.[20] In fact, Bronowski and his assistant, Stu Ross, were actively involved in discussions with the junior scientists to find out whether their research activities could be incorporated into the CBHA. Eventually Bruno began informal discussions with Theodore (Ted) Friedmann, an M.D. interested in genetic diseases who had extensive research training at the University of Cambridge and the National Institutes of Health (NIH).[21] He arrived at the Salk in September 1968 to spend a year in Dulbecco's lab studying the structure of some viruses with the idea of making virus particles useful for medicine. After that year he remained in La Jolla and took a faculty position in the

Pediatrics Department of the new UCSD School of Medicine. Since his space at the medical school was not yet ready, he continued to work in borrowed space in Holley's lab at the Salk and to interact with Bruno and the CBHA while having a faculty appointment at UCSD.

Friedmann wanted to apply genetic technology to medical problems. One of his early interests was the use of genetic screening technologies to detect disease. Another of his interests was attempting to engineer artificial viruses that would carry bits of foreign DNA to achieve the genetic modification of cells. Originally Friedmann did not have any training or particular interest in policy or in nonscientific issues. It is only when he became involved with Bruno and other members of the CBHA that he discovered his interest in questions such as what to do with genetic information and how to use that information as a doctor to serve the public.

Friedmann was working in a lab near other people who were concerned about the same issues. One of those people was Richard (Dick) Roblin. Unlike Friedmann, Roblin had a longtime interest in political and ethical questions, and his liberal arts education included an excellent course in political science at Williams College, in Williamstown, Massachusetts. Like Ted, Dick was fascinated with the possibility of new gene-based treatments for some human genetic diseases. They both became involved with questions of bioethics—which at the time had not yet been established as a discipline—when they questioned the risks of applying these new therapies to humans.

Both men became active members of the CBHA Commission on Biology, Ethics, and the Law, writing white papers and making presentations that were discussed at commission meetings with experts in nonscientific areas, including Abram Chayes, professor of law at Harvard, and David L. Bazelon, one of the Salk Institute trustees and chief judge of the U.S. Court of Appeals of the District of Columbia. Those discussions resulted in two landmark papers that, with the help of Bruno, were published in prestigious magazines reaching a wide audience, *Scientific American* and *Science*.[22]

Those papers in particular had a great impact on the rest of their careers. The papers attracted the attention of Paul Berg, who had met Dick Roblin briefly while on sabbatical at the Salk Institute. This led to Roblin's involvement in the recombinant DNA policy debates, the famous 1974 moratorium letter, and his becoming a member of the Organizing Committee for the 1975 Asilomar Conference on Recombinant DNA.[23] Upon returning to the East Coast to teach and do research at Harvard Medical School, Roblin continued to explore the bioethical implications of developments in genetics and molecular biology. In 2001, when he retired from scientific research, he became the scientific director of the President's Council on Bioethics.

As for Ted Friedmann, he pursued his interests in bioethics throughout his distinguished career at UCSD School of Medicine, where he holds the Whitehill Chair in Biomedical Ethics and serves as director of the Gene Therapy Lab of the Department of Pediatrics.[24] In 2002 he was named chair of the Recombinant DNA Advisory Committee at the NIH, which oversees federally funded research involving recombinant DNA.

A visitor and seminar speaker who had considerable influence on some of the junior scientists at the institute was John R. Platt. Bronowski likely invited Platt to the institute because, at the end of 1969, he had published an article in *Science* that struck a chord with many people in the sociopolitical mood of the late 1960s. Platt's article, titled "What We Must Do," warned about imminent and acute worldwide crises that threatened the human race.[25] These included nuclear escalation, famine, overpopulation, political crises and protests, pollution, poverty, and, of course, disease. He called for the large-scale mobilization of scientists and the formation of research and task forces not unlike those set up in wartime. The first step would be to organize technical discussion and education groups in every laboratory. Each of these small study groups would have its own topic that it would analyze in depth.

The reaction at the Salk Institute was amazing.[26] Human health, of course, was the area that Salk Institute scientists were best suited to study because of their biology backgrounds, and a health study group was formed. That group included four junior scientists that made life-changing decisions as a result of those discussions. Two of the four, Donato Cioli and Paul M. Knopf, worked in the Lennox lab, while the other two, F. Alan Sher and Italo M. Cesari, worked in the Cohn lab. The CBHA was there to support the efforts of the health study group by helping them to collect relevant literature. Bronowski, Boardman, and Ross, as well as Lennox and Cohn, attended some of those group meetings. Eventually the group was made aware of a review article entitled "The Unconquered Plague."[27] It was about schistosomiasis, a disease caused by a parasitic worm that at the time affected at least two hundred million people in developing countries. Cioli was assigned the task of researching the disease and reporting his findings to the group. His presentation of the problem was very convincing. The disease was understudied, the biological system was fascinating, and the crisis was

FIGURE 14. Rockefeller Foundation meeting on schistosomiasis, Lake Como, 1972. From left to right: Alan Sher, Donato Cioli, Paul Knopf, Italo Cesari. Courtesy Alan Sher.

highly relevant to the well-being of mankind. Cioli did such a great job that the four members of the group mentioned above decided to adopt the disease as their research area when the time came to set up their independent laboratories.

Indeed, all four became parasitologists, and each contributed to the understanding of that dreadful disease. Cioli returned to Rome, where he set up his schistosoma research group at the Institute of Cell Biology, while Cesari returned home to Venezuela to establish a schistosoma biochemistry laboratory at the Venezuelan Institute of Scientific Investigation in Caracas. Knopf and Sher took positions in the United States. Knopf accepted a faculty appointment at Brown University in Providence, Rhode Island, where he eventually became chairman of the Department of Molecular Microbiology and Immunology. Sher took a research position at the National Institutes of Health in Bethesda, Maryland, where he is now chief of the Laboratory of Parasitic Diseases at the National Institute of Allergy and Infectious Diseases. By 1972 all four had left the Salk Institute, but they were briefly reunited in April of that year in Italy, on Lake Como, for a meeting on the immunological control of schistosomiasis sponsored by the Rockefeller Foundation.

Weekly seminars known as Speaking-on-Tuesday were yet another activity organized by junior scientists and encouraged by the CBHA. These were informal talks, primarily by visitors, junior scientists, and members of the community, about various topics of general interest, such as pollution, deep sea drilling, the analysis of moon rocks, and Indian reservations and schools in San Diego County. Donato Cioli, who was in charge of organizing those seminars in 1969, remembered a very special presentation by a visiting French scholar, Edgar Morin. In his broken but uninhibited English Morin spoke about the May 1968 student revolt in Paris, a topic that clearly resonated with the audience at a time when protest was everywhere.[28]

Morin spent six months at the Salk Institute, from September 1969 to March 1970.[29] A French philosopher and sociologist—*un penseur* (a thinker)—living in Paris, he was well known to John Hunt and to Jacques Monod, who had recommended him.[30] At the Salk, Morin participated in symposia on biology and the humanities and actively interacted with the scientists. Inviting Morin for a lengthy stay at a top biological research institute was to be, according to John Hunt, an experiment, and it was successful beyond expectations.[31] Curious about everything, Morin had an inquiring mind, was industrious, enjoyed reading and talking to people, and worked at understanding the science. He was like a student again. After he left he wrote a lighthearted book about his experiences in California.[32] His *Journal de Californie,* dedicated to John Hunt, Jacques Monod, and Jonas Salk, was a success in France because it was entertaining and timely. People in France were interested in California of the 1960s, a new world that they wanted to know all about and to visit. More importantly, however, Edgar's stay at the Salk Institute had a great impact on him. In his own words, "The year at the Salk was a rebirth for me."[33] What he learned in La Jolla became for him the basis for years of intellectual productivity.

By the summer of 1970 Morin and Hunt were back in Paris. Eventually Hunt was approached by Philippe Daudy, the son-in-law of Henry Gouin, the man whose family owned the Abbey of Royaumont, some twenty miles north of Paris, and had set up the Fondation Royaumont in 1964.[34] Daudy informed Hunt that Royaumont was looking to add a new dimension to its activities to complement their distinguished programs that were primarily in music. Hunt thought of his friends and fellow conspirators in Paris, Morin and Monod. They agreed with Hunt's idea to follow the impulse of bringing the humanities and the sciences together in a European setting with European financing, similar to the CBHA at the Salk. The hosting of colloquia by the Fondation Royaumont and the use of the beautiful abbey as a venue made the plan especially attractive. With some funding from the philanthropist Cyrus Eaton, the initiator of the Pugwash movement, the French group organized its first colloquium in September 1972 to test the idea. Morin and Massimo Piattelli, a student of Monod, were co-organizers of the colloquium L'Unité de L'Homme and coeditors of the book based on the contributions of the participants.[35] The meeting and book were such a success that they resulted in the creation of a more permanent organization, the Centre Royaumont pour une Science de l'Homme.[36]

What Monod, Morin, and Hunt did at Royaumont is what they would have liked to have seen done at the Salk Institute. Clearly, the impetus for their efforts at Royaumont came from the initial effort at the Salk, and, indeed, many of the same characters were involved, including Salva Luria. The connection was very clear.[37] In France, however, Monod was the inspiration and he imbued the Centre Royaumont with a more philosophical mentality and greater intellectual and social ambitions than the modest social science project of the Salk Institute.

Nonetheless, the activities of the CBHA had a positive effect at the Salk Institute. The involvement of the junior scientists expanded their horizons and in some cases changed their lives. All of those who were engaged in the CBHA's programs learned and benefited from their interactions. They were buoyed by the activities of the CBHA and proud to be involved in pursuits that were unique for a biological research institute.

CHAPTER 12

A Napoleon from Byzantium

Wir danken's unserem Führer.[1]

—Erwin Schrödinger

The takeover of the institute presidency by Frederic (Fred) de Hoffmann in 1972 (see chapter 10) represented the end of Salk's original vision for our institute. The institute changed from idealistic to realistic and from lean to mean. De Hoffmann's style came as a shock. Like Byzantine politics, all of his dealings were complicated, underhanded, and non-negotiable. De Hoffmann greatly admired Napoleon, whose life he had studied as a boy, and the experience of the Emperor of the French probably encouraged Fred's delusions of grandeur and thirst for absolute power.[2] He preferred to keep his family background a mystery. Most who knew him did not even know his original nationality or family name. However, his background interestingly reflects the tragic and complicated times in Europe and especially the plight of the Jews.

Frederic de Hoffmann was born Fritz Hoffmann in 1924 in Vienna, Austria.[3] His father, Otto, was also born in Vienna, but he used the name Otto Hoffmann von Vagujhely, after a town in Hungary.[4] Fritz's mother, born Marianne Halphen, was a Czech born in Prague. Both of Fritz's parents were Jewish, but his mother had taken the usual— although useless—precautions to protect her child from anti-Semitism. She renounced the Jewish faith shortly before Fritz was born, and his birth certificate specified that he was baptized by a priest. It appears that Fritz's mother died soon after his birth, as he lived with his father in Vienna and was raised by a nanny. In 1933, when Fritz was nine years old, his father died and he moved to Prague to live with his maternal uncle Francis Halphen, an attorney, and his aunt, Katerina. That is where Fritz attended high school in German.

In March 1939, when Hitler marched into Czechoslovakia, it was time to plan the family's escape to America. Uncle Francis immediately sent Fritz to England with a Hungarian passport issued in Prague. Fritz enrolled in high school in London to complete his secondary education and to obtain the certificate that would allow him to apply to an American university. Meanwhile, in Prague, his uncle Francis was collecting affidavits of support from relatives and friends living in the United States, which were essential for them to immigrate. However, their plans went awry when, on September 1, Hitler attacked Poland. Two days later, as Britain and France declared war on Germany, World War II began (see prologue).

Francis and Katerina Halphen eventually left Prague, and by January 1940 Francis held a position in Paris at the Czech National Committee, a de facto Czechoslovak government-in-exile. Presumably the plan was for Fritz and his family to be reunited, perhaps in France or in England, and then sail to America together, but this did not occur. By May 1940, as Nazi Germany invaded Belgium and entered France, the Halphens fled Paris. It took them several months to reach Lisbon, Portugal, where they were able to catch one of the last crossings of the Greek ship the SS *Nea Hellas*, which arrived in New York on October 13, 1940.

It seems appropriate to note that the SS *Nea Hellas* was the same ship that had brought Salva Luria to New York one month earlier, on September 12 (see chapter 4). That brave ship sailed between Piraeus (Greece), Lisbon, and New York during most of 1940, saving the lives of thousands of refugees from Europe. As Fascist Italy invaded Greece on October 28, all crossings to America were cancelled, and the SS *Nea Hellas* did service as a troop transport for the Allies.[5]

Meanwhile, Fritz was stranded in England, where the Blitz on London had begun in September. He was evacuated to the country and took up residence near Cardiff, Wales. However, in order to organize his trip to America he still had to travel to London, where it was necessary to visit the Royal Hungarian Legation to obtain a certificate of Hungarian citizenship and a statement of good conduct, which were required to apply for a U.S. immigration visa. At the American consulate in London, he filled out a visa application dated January 1941, which stated his intention to join his uncle in New York, to apply for residence, and to remain permanently. What was revealing in this application was that his port of embarkation was "uncertain." He obtained a

U.S. visa valid until May, and his Hungarian passport was valid until June. All he needed was a ship—soon.

Eventually he managed to get passage from Cardiff on a Greek freighter, the SS *Mount Kyllene.* and disembarked in Norfolk, Virginia, on March 5, 1941. He was seventeen and the only passenger on that freighter. He soon moved to Cambridge, Massachusetts, where he started his studies in physics at Harvard. Before he completed his degree, he was picked out of Harvard and taken to Los Alamos to join the Manhattan Project as a civilian. When he arrived at Los Alamos in early January 1944, he was assigned to the Theoretical Division headed by Hans Bethe.[6] At Los Alamos he met a young physicist who was to become one of the founders of the Salk Institute, Edwin Lennox (see chapter 6). During this period Fritz made contacts with many world-renowned physicists, including Enrico Fermi, Richard Feynman, Edward Teller, Niels Bohr, and, of course, J. Robert Oppenheimer. After the war Fritz returned to Harvard, and he completed his Ph.D. in 1948. He then became a staff member of the Atomic Energy Commission (AEC). While at Harvard he had obtained American citizenship and changed his name from Fritz Albert Wilhelm Hoffmann de Vagujhely to Frederic (Fred) de Hoffmann.

In early 1949 de Hoffmann returned to Los Alamos, where he became the protégé and factotum of Teller, who eventually found Fred to be an indispensable assistant. For some years Teller's pet project had been the hydrogen bomb, or "Super," but it had been abandoned during the Manhattan Project in favor of the atomic bomb. The "Super" was a controversial project, and many, including Oppenheimer, believed that a bomb some hundreds of times more powerful than the atomic bomb was neither needed nor desirable. De Hoffmann, like Teller, disagreed and wanted to continue to explore the idea. Then, in August 1949, the Soviets exploded their first atomic bomb, and the arms race was on. By January 1950 President Truman had directed the AEC to continue work on the hydrogen bomb. Oppenheimer's opposition to the "Super" was to be his downfall, as it was one reason that Teller testified against him (see chapter 2).[7] De Hoffmann, on the other hand, became Teller's deputy at the AEC and accompanied him to important conferences on nuclear affairs.[8]

In August 1955, de Hoffmann attended the first Atoms for Peace conference in Geneva, the first meeting between nuclear physicists of the West and East. As an AEC representative, de Hoffmann was a member of an international team of scientific secretaries appointed to determine what information would be exchanged. In particular, they would decide what should be kept secret and what could be safely disclosed, such as the design of commercial reactors. Representatives of industry were observing, and General Dynamics (GD), as the largest defense contractor in the United States, was most interested. The Convair Astronautics Division of GD in San Diego was already building the first operational intercontinental ballistic missile, the Atlas.[9] John Jay Hopkins, the chairman of GD, thought that the time had come to enter the nuclear energy business and had asked Teller for the name of an expert. Not surprisingly, Teller recommended Fred de Hoffmann.[10]

De Hoffmann moved to San Diego as a vice-president of GD. With Hopkins he started planning the development of peaceful applications of atomic energy. In 1955 they founded a new division of GD, General Atomic (GA), with de Hoffmann as general manager. GA obtained funding and 320 acres of pueblo land on the Torrey Pines Mesa, where they built a lavish complex that included a circular technical library, a cafeteria, tennis courts, and a swimming pool. Called the John Jay Hopkins Laboratory for Pure and Applied Science, the complex was dedicated by Niels Bohr in 1959 (see chapter 5).[11] Leo Szilard had spent some time at GA in 1959 and had then insisted that Salk visit La Jolla, which he did in August of that year. On a subsequent trip to La Jolla with O'Connor, Salk had visited the GA building and eventually met de Hoffmann, who had been very welcoming and supportive at the time of the June 1960 ballot.[12] Indeed, the presence and early success of GA had added to the appeal of La Jolla as a site for the Salk Institute.

Inspired by the first Geneva Atoms for Peace conference, GA had quickly designed TRIGA, a small and safe nuclear reactor meant to prepare isotopes. GA had even sent a TRIGA prototype to the second Geneva conference in 1958.[13] Teller claimed credit for the idea of building safe nuclear reactors, but TRIGA was based on a clever scheme by Freeman Dyson. It certainly was an impressive start, but de Hoffmann had bigger plans. He bet heavily on building a foolproof big power reactor called HTGR, although it never sold and was a loss. Moreover, after the launch of Sputnik in 1957, the space race was on, and de Hoffmann started dreaming of building an interplanetary atomic spaceship, Orion.[14] It was not to be just any spaceship but a four-thousand-ton vehicle propelled by nuclear bombs (preferably "Super") and derived from an original concept by the mathematician Stanislaw Ulam at Los Alamos. From 1957 to 1965, Project Orion was ongoing at GA, where de Hoffmann was trying to re-create the excitement of Los Alamos.

Enthusiasm was high, and the physicists' motto was "Saturn by 1970." After some seven years of effort by a team of the best minds with almost unlimited money, Orion never materialized. The 1963 Test Ban Treaty against nuclear explosions in the atmosphere and in space made Orion outdated before it had a chance to exist.

In 1967, General Dynamics sold GA to Gulf Oil, and soon support for de Hoffmann's grandiose plans ran out. By 1969 de Hoffmann had been laid off with a golden parachute and was looking for a job. Around this time Slater announced his imminent resignation to assume the presidency of the Aspen Institute (see chapter 10). Since de Hoffmann and Salk knew each other and both lived in La Jolla, it is easy to imagine that they met and talked about their respective predicaments. Perhaps de Hoffmann might consider a temporary position as chancellor of the Salk Institute until the Slater situation—and the resulting financial crisis—was resolved. By May 1970 Slater had proposed de Hoffmann as chancellor, and in August he was appointed chancellor and CEO of the institute.

At the time this seemed like an ideal arrangement. It was to be temporary, and de Hoffmann already resided in a fancy house he had built on Black Estate property adjacent to the Salk Institute in 1961. It was not a big commitment for either party, as de Hoffmann might still have hoped to find a position in a field more closely related to physics, and Salk was certainly hoping that the Slater crisis would soon be amicably resolved. However, this was not to be, and de Hoffmann was elected president of the Salk Institute in February 1972.

At General Atomic, de Hoffmann was king of the realm. He had built the facility and handpicked the staff and advisors. Nothing happened that he did not know about immediately; he was on top of the division's scientific and technical challenges and in total control. Moreover, as long as GA was a division of General Dynamics, he essentially had a blank check. His position at the Salk Institute was quite different. He was to take over an institute that had been built by someone else; he had to deal with staff, scientists, and advisors that he did not know; he did not understand the biological sciences; and the institute was broke. Somehow he needed to gain control and he was not ready to share the power with anyone.

Most threatening to de Hoffmann's authority was, of course, the institute's director, Jonas Salk. De Hoffmann's first move was to neutralize Jonas's power as much as possible. Since he was the only candidate for the president's job, de Hoffmann was in a position to negotiate

FIGURE 15. Fellows and President Fred de Hoffmann, 1974. From left to right: Jacques Monod, Gerry Edelman, Mel Cohn, Leslie Orgel, Bob Holley, Jacob Bronowski, Fred de Hoffmann, Salva Luria, Paul Berg, Ed Lennox, Roger Guillemin. Note the absence of Jonas Salk. Courtesy The Salk Institute.

the terms, and he had stipulated as a condition of his taking the job that Jonas Salk relinquish to him all of his previous duties and responsibilities as director. Eventually, in 1975, an agreement was reached according to which Jonas Salk was awarded the title of founding director, a title that carried some privileges but no power. What de Hoffmann wanted more than anything was to remove the word "Salk" from the name of the institute, but in that he failed. Today (2012) it is still called the Salk Institute.

De Hoffmann systematically worked at weakening the board of trustees, the most powerful body at the institute. When commanding chairman McCloy resigned in 1974, he was replaced by Samuel B. Stewart, a friend and a supporter of de Hoffmann who was an executive at Bank of America. Carefully, de Hoffmann assembled a board that was eager to ratify all of his wishes, content with receiving little information about the institute, and appreciated that meetings—which were usually held off campus to minimize contact between trustees and the faculty and executive staff—were both rare and brief. The board members were not expected to play a role in fund raising or to contribute any financial support themselves. The board became nongoverning and nongiving, just as de Hoffmann wanted it. He did not want a generous board, because that would make for powerful trustees. The function of the board was vague, purely a formality, and there was no limit on the length of service of the trustees. Moreover, most of the trustees were from the East Coast, as the San Diego community still included too many friends of Jonas.

As far as the nonresident Fellows were concerned, he kept them at a distance. They were no longer encouraged to spend extended periods at the institute and to visit the labs. De Hoffmann made sure that the group of nonresident Fellows always included some of his supporters who were aware of his wishes and would help deliver the desired outcome. He also avoided face-to-face meetings and discussions as much as possible. He kept the nonresident Fellows informed through newsletters and polled them by phone. His correspondence with the nonresident faculty communicated only selected news and always demanded the greatest secrecy.[15] Occasionally de Hoffmann was caught in his web of lies, misrepresenting the position of one person to another. However, most of the nonresident Fellows put up with his schemes.

Handling the resident faculty was a somewhat trickier task. De Hoffmann decided to deal with the resident faculty through a single faculty member who served as a liaison and was designated by him in perpetuity. He also handpicked the faculty members who served on the board as faculty representatives to ensure that they would not start any trouble. It soon became very clear who de Hoffmann's protégés were, and this divided and weakened the faculty. Moreover, the faculty members who refused to machinate with him were marginalized and weakened as de Hoffmann ignored their demands or even divided their laboratories. After the death of Bronowski in 1974, the Council for Biology in Human Affairs was dismantled, the Szilard Memorial Collection of rare books was sold, and all the intellectual activities that Bruno had inspired were terminated (see chapter 11).

Since de Hoffmann refused to delegate authority, he performed many functions himself with the help of a staff following strict orders. These functions included policy making, academic decisions, allocation of resources, communication, and, of course, fund raising. To deal with the financial crisis, de Hoffmann used the business approach: cutting costs, downsizing, layoffs, and austerity. He fully expected that the faculty would make sacrifices and increase the funds they acquired through government grants. He never raised a sizeable endowment. And, to further undermine Jonas's influence, he chose not to engage the San Diego community and raised funds primarily in New York and established many contacts in Europe.

In the early 1970s three of the founders left the institute. Without resigning their position as Fellows, Dulbecco and Lennox took leaves of absence that extended to several years to work in England. In 1973, after two six-year terms as a nonresident Fellow, Crick did not renew his appointment. Jonas had recruited all three of them in the early 1960s. Following a Machiavellian plan, de Hoffmann attracted Dulbecco, Lennox, and Crick back to the institute on his own terms and made them his allies. De Hoffmann profoundly hurt Jonas, who kept a stiff upper lip, studied French, and became increasingly involved with the exciting world of his wife, Françoise Gilot. Still, Jonas always remained ready to help when his institute needed him. There is no doubt that the construction of the institute's East Building, which was fraught with controversy, would not have been possible without Jonas's involvement in the early 1990s.

De Hoffmann had embedded himself in the Salk Institute so craftily that he remained in power for eighteen years. He resigned in 1988 when it was discovered that he had contracted AIDS as a result of a blood transfusion. His last instructions to his staff were to destroy all the administrative archives of his tenure.

It must be said, however, that three of de Hoffmann's achievements turned out to be very constructive for the Salk Institute in spite of having been done for the wrong reasons. Encouraging the split of the Neuroendocrinology Laboratory was clearly done to weaken the power of Roger Guillemin, whose influence had increased after he was awarded the Nobel Prize in 1977. In 1978 de Hoffmann arranged for a group of scientists working in the Guillemin lab to set up a competing peptide biology laboratory right on the Salk campus. That group was very successful on its own, and its presence at the institute was eventually a positive development, although it was created in a way that was disrespectful and painful not only to Guillemin but also to the rest of the faculty. The introduction of plant biology at the Salk Institute was another success that was achieved by de Hoffmann in an unpleasant and underhanded way. The faculty had concerns because it was linked—through a complicated scheme—to floating SIBIA, the Salk Institute Biotechnology/Industrial Associates, a private biotech company that was an affiliate of the Salk Institute. SIBIA was created in 1981 without the knowledge of most members of the faculty, and there were rumors of conflict of interest. Eventually, however, the institute was lucky because SIBIA introduced an outstanding plant biologist, Chris Lamb, who founded the distinguished Plant Biology Laboratory. Finally, attracting Francis Crick back to La Jolla in 1977 was certainly very positive, as Crick furthered the development of the neurosciences—and the area of vision research in particular—at the institute, where he remained active until his death in 2004.

The de Hoffmann era was, however, a dark period for many at the Salk Institute. His legacy was a weak board, a divided faculty, and little endowment. However, the heritage of the founders gave the institute the strength to survive. That heritage was scientific excellence in several directions that mostly remained as originally planned including neurobiology (from brain hormones to language), aging and gene regulation (in normal cells and cancerous ones), and cell differentiation (from immune cells to stem cells). Although the de Hoffmann regime might be viewed as the end of the genesis of the Salk Institute, our institute grew primarily because it had assets more valuable than money: a reputation of distinction, a great building, and many loyal employees, both faculty and support staff, for whom the Salk Institute was not just a job, it was a dream.

Dreams die hard.

Epilogue

Fifty Years Later

The only constant is change.[1]

Today (2012), the Salk Institute scientific staff numbers about 850, some fifty-five of whom are resident faculty members. It is still a small stand-alone biological research institute, but in fifty years it has grown to about twice the size of 450 originally planned by Jonas Salk.[2] The scientific excellence that characterized the founders established the standard by which the next generation of the institute's scientists were selected. Although it is still small, the institute has lost the feel of a family where everybody knows your name and the cachet of a boutique where every product is special.

Growth of the institute has been slow, limited by available funds. However, one change that happened quickly was the replacement of the title of "Fellow." The second-generation faculty members wanted to be called "professors," as they would be at a university. (Amusingly, however, the room in the Kahn building where the founding Fellows used to meet is still called "the Fellows' Room.") Some wonder why senior professors were once called Fellows. That title goes back to a time when the Salk Institute only existed on paper and a handful of risk-taking dreamers resigned their safe faculty positions to live on precarious fellowships while working to bring our institute into existence. Since two of the founders—Bronowski and Crick—were British, this title was very much in keeping with their tradition of having senior academics as Fellows attached to colleges.

The nonresident faculty members are still called "nonresident Fellows." In fact, today's nonresident Fellows—still a very distinguished group—are too busy to spend more than one or two days a year at the institute. They still vote on promotions and appointments, but now they have little influence on the long-term scientific direction of the institute. The annual Fellows colloquium, originally designed to explore possible areas for future expansion, has been abandoned for lack of time. This is regrettable because today's nonresident Fellows could continue to make important contributions to the scientific future of the institute, since they enlarge the range of expertise of our resident faculty. The interests of our institute used to be broader: origins of life (see chapter 9), history of science, and biology in human affairs (see chapter 11) are now forgotten.

In the early days, the group of resident Fellows was small enough that the position of chairman of the faculty was simply rotated, and junior faculty ranks did not exist. Under de Hoffmann, the faculty was fictive and powerless. In reaction to the de Hoffmann dictatorship, the Academic Council was created. Acting as the elected representative of the faculty, it includes junior faculty members and an elected chairman. The Academic Council is the main body formulating academic policies, but, in principle, the full faculty, which meets quarterly, retains ultimate academic authority. Following the early philosophy of "sitting on both sides of the fence," our current institute president, William Brody, is also a professor at the institute and belongs to both the faculty and the administration. No one can deny the growing influence on institute policies of an administration that now includes a number of vice-presidents.

Our present board of trustees, which includes prominent members of the San Diego community, is generous, involved, and powerful. The chair is Irwin Jacobs, a professor of engineering at UCSD, founder of the cell phone giant Qualcomm, and philanthropist. The March of Dimes continues to support our institute handsomely, and its president, Jennifer L. Howse, is a leading member of our board. Our teaching relationship with UCSD has been maintained; most Salk faculty members are adjunct professors at the university, do some teaching, and train graduate students. The institute faculty has discussed, but always rejected, the idea of becoming a degree-granting institution.

Our institute is, however, caught up in the problems and trends of the times. Biology has become big science, especially with the explosion of genome sequencing, which has been driven by improved technology and generates data faster than it can be understood. Biological research has become less individual and more often a group effort. The Salk

Institute is home to a number of research centers at which investigators with an affinity can interact, share resources, and join forces to obtain financial support. As it has become more difficult for individual investigators to raise funds, these centers offer new opportunities to acquire funds, first to set up research centers and then to support program projects and core grants.

The National Institutes of Health (NIH) now favors applying available knowledge rather than generating new knowledge. For example, the NIH wants to know *which* drugs work, not *why* they work. The development of technology that allows the fast screening of a large number of drugs provides an approach that does not require understanding. Applied research has become translational research in which active steps are taken to find practical applications of basic research discoveries. To illustrate that point, the NIH has created the National Center for Advancing Translational Sciences (NCATS), and the Salk Institute Office of Technology Management has changed its name to Office of Technology Development. The funding policies of the government today discourage basic research driven by curiosity. As the president of the Max Planck Society recently pointed out, this policy is shortsighted.[3] Because basic research often takes years to lead to applications, it is too risky for businesses to finance it. It is up to governments, then, to invest in the basic research upon which our future depends.

Jonas was well aware that the development of his institute would be an evolutionary process, and he viewed it as an experiment. The experiment continues today, and in these changing times the Salk Institute is evolving and adapting as well. It has the brains that made it a scientific success story, and it has the inspirational building that made it a historical landmark. Indeed, the Kahn building has endured as a symbol of Salk's original dream, and perhaps that was its major purpose: to serve as a reminder of an enterprise of great pitch and moment. Jonas's institute did not turn out exactly as he had hoped, and some of his early ideas were abandoned, such as the creation of a small community of scholars, enriched by a constantly renewed group of visitors, that would bridge the sciences and the humanities. However, as it stands the institute is a remarkable legacy.

Our institute has attracted the attention of many distinguished scientists, among them a number of Nobel laureates who have joined the faculty as resident, nonresident, or visiting members. It has also trained a school of junior researchers, many of whom have won major awards, including Nobel, Lasker, and Wolf prizes. Its excellence has been recognized by very high rankings in many objective surveys. The success of our institute proves that in this era of big science there is a role for a small institute without boundaries: this turns out to be a strategic advantage, as the close interaction of scientists across disciplines permits unique insights and discoveries.

Nevertheless, the institute would not have come into being if it were not for the resilience of Jonas to the pressures of high risk and rejection. He put his faith in Basil O'Connor and the support of the March of Dimes and risked his reputation by making promises he could not be sure he would be able to keep. He risked his pride by daring to attract a team of founding scholars—scientists, trustees, and administrators—who were widely regarded as more accomplished. He was demeaned, abandoned, and hurt by many, from Sabin to Slater and de Hoffmann, to name only a few. But Jonas's spirit was strong and could not be broken. He remained undaunted and never complained. He was infinitely optimistic, patient, and kind, and he was a caring friend to many.

On June 23, 1995, Jonas Salk died in La Jolla. The institute that bears his name remains as his second outstanding achievement.

Notes

PREFACE

1. Carter 1966.

PROLOGUE: THE GREATEST GENERATION

1. The prologue title and FDR quote have been used by Tom Brokaw (1998).

2. See, e.g., Carter 1966; Oshinsky 2005; Paul 1971; Smith 1990.

3. Dulbecco 1990.

4. MCP.

5. Lennox, interview by the author, June 2008, SBP.

6. JBP.

7. Weaver 1970b.

8. Lanouette 1992.

9. Debré 1996; Soulier 1997.

10. Olby 2009.

11. "It's war!"

12. For book-length accounts of the Manhattan Project, see, for example, Rhodes 1986; Bird and Sherwin 2006.

13. Bethe eventually became Lennox's Ph.D. thesis advisor and received the Nobel Prize in Physics in 1967.

14. MCP and Melvin Cohn, interview by the author, SBP.

15. Bronowski 1959.

16. Crick 1988.

17. Dulbecco 1990, 85.

1. BEFORE AND AFTER ANN ARBOR

1. "Genius is but a great talent for patience."

2. Polio is primarily a gastrointestinal viral infection. It was not rare but caused neurological problems in only one out of hundreds of patients.

3. Carter 1966; Oshinsky 2005; Paul 1971; Smith 1990.

4. Quote from Edward R. Murrow.

5. Today the March of Dimes Birth Defects Foundation.

6. Biographical details about O'Connor are from Fishbein 1957 and Takaro 2004. The relationship between Salk and O'Connor as friends and partners is described in a forthcoming article by David Rose, the archivist of the MOD.

7. After 1934 O'Connor became senior partner at the law firm of O'Connor and Farber.

8. As of 2011, FDR's summer house is the center of the Roosevelt-Campobello International Park (www.fdr.net).

9. Fear of prejudice compelled FDR to hide his condition in the 1920s.

10. As of 2011 Georgia Warm Springs was a state-managed rehab facility, the Roosevelt Warm Springs Institute for Rehabilitation (www.rooseveltrehab.org).

11. Sills 1980.

12. Benison 1967.

13. The role of science and technology, epitomized by the atomic bomb, in winning World War II triggered U.S. government support for scientific research.

14. For example, the NFIP sponsored the famous 1953 Cold Spring Harbor symposium on viruses, the meeting at which Jim Watson presented the structure of DNA.

15. He introduced "indirect costs" and the requirement for progress reports.

16. Paul 1974.

17. Passage of a virus through animals or through tissue culture cells selects for rare mutant strains less virulent to humans.

18. Paull 1986.

19. Until then it was widely believed that the virus entered the body through the nose, did not enter the bloodstream, and grew only in nerve cells.

20. This won them the 1954 Nobel Prize in Medicine.

21. Carter 1966.

22. The full text is in Carter 1966.

23. Indeed, Sabin's vaccine was not licensed until 1962.

24. Harry Weaver became director of research at the American Cancer Society and later at the National Multiple Sclerosis Society.

25. Paul 1971.

26. Offit 2005.

27. Scudellari 2010.

28. Polio eradication remains a challenge. Roberts 2010; Larson and Ghinai 2011.

29. Goldman et al. 2003.

2. DOCTOR POLIO MEETS DOCTOR ATOMIC

1. Transcript of the February 22, 1958, Advisory Committee meeting, JSP b345/f1.

2. Alberts 1986; Paull 1986.

3. During the June 1948 blockade of West Berlin, Clay ordered and organized the Berlin airlift of food and supplies and maintained it for almost one year. He is remembered as the hero who saved Berliners from starvation.

4. Solow 1958.

5. JSP b344/f1.

6. Salk did not use this modern expression, but it conveys his meaning.

7. Batterson 2006; Regis 1987.

8. Manley 1974.

9. Thorpe and Shapin 2000.

10. Cited in Thorpe and Shapin 2000.

11. Many books and articles have been written about the Manhattan Project and Oppenheimer's life. See, for example, Rhodes 1986; Teller 2001; Cassidy 2005; Herken 2002; Scherer and Fletcher 2008; Kelly 2006.

12. Regis 1987.

13. Salk and Oppenheimer had actually become acquainted in Pittsburgh in 1953. That year Salk and his family had moved from the suburbs to the city of Pittsburgh. Quite by chance their next-door neighbor was an aunt of Oppenheimer, who visited her regularly from Princeton. This was also the time when Salk's first positive results about the polio vaccine had been reported by the press (see chapter 1). Oppenheimer had a personal interest in polio because in 1951 his daughter Toni (age seven) had suffered a mild case of the disease. On occasions of visits to his aunt Oppenheimer also visited the Salk home. Peter Salk, interview by the author, March 2006, and Françoise Gilot-Salk, conversations with the author, August 2012, SBP.

14. JSP b345/f1.

15. JSP b345/f4.

16. JSP b345/f1.

17. This must have been November or December 1956, shortly after Salk first met Szilard. See chapter 3.

18. A draft dated March 14, 1958, is available in JSP b344/f3.

19. The organizational features of the National Foundation–March of Dimes, its evolution, and the basis for its success have been well analyzed by Sills 1980.

20. The materials concerning the Institute for Advanced Study sent by Salk to O'Connor included the "Report of the Director 1948–1953," written by Oppenheimer in 1954 and found in JSP b146/f1.

21. Batterson 2006; Regis 1987; Flexner 1960.

22. Flexner 1960, 236.

23. Salk to Oppenheimer, January 13, 1959, JSP b349/f7.

24. Salk to Oppenheimer, February 7, 1959, JSP b349/f7.

25. Heuck 1994.

26. Nelson 1968. Salk Hall was eventually remodeled to house the schools of dentistry and pharmacy.

3. ENTER LEO SZILARD

Many of the details reported by Lanouette (1992) are from the three volumes edited by Szilard's wife, Gertrud (Trude) Weiss Szilard. Trude contributed much to her husband's legacy by having his scientific and professional papers edited and published. She collected his personal papers and letters, and her brother Egon Weiss released them in 1983, two years after her death. Feld and Szilard 1972; Weart and Szilard 1978; Hawkins, Greb, and Szilard 1987.

1. Lanouette 1992, 85.

2. See chapter 2. Salk to Oppenheimer, January 13, 1959, JSP b349/f7.

3. See chapter 2. Salk to Oppenheimer, February 7, 1959, JSP b349/f7.

4. Salk to O'Connor, September 26, 1958, JSP b349/f7.

5. Lanouette 1992.

6. Lanouette's book shows a photograph of the Vidor Villa, which became the dormitory for students at the Béla Bartók College of Music.

7. At the time there existed two types of high schools in Hungary: gymnasiums covering classical subjects and practical schools teaching science and technology.

8. Lanouette 1992, 392. While a consultant for the Conservation Foundation (see chapter 4) Szilard so humiliated a gynecologist who erroneously claimed that hesperidin was a birth control drug that the doctor committed suicide.

9. Many Jews living in central Europe at the time renounced the Jewish faith: the Wigners became Lutherans and the von Neumanns Roman Catholics.

10. After World War II the Kaiser Wilhelm Society was threatened with closure because of its connections to the Nazi regime. This was avoided when Max Planck took over the temporary presidency of the society in May 1945. Planck was respected by all not only as the founder of quantum physics, but also for his outspoken opposition to Hitler. Eventually Planck consented to lend his name to the society, and a new organization, the Max Planck Society for the Advancement of Science, was founded in 1948 in Göttingen. From the start, the institutes created by the Max Planck Society were different from the Kaiser Wilhelm institutes in that they were entirely dependent on public funding and focused on basic research. For the history of the Max Planck Society, see www.mpg.de.

11. Lanouette 1992, 74.

12. Lanouette 1992, 133.

13. Meitner and Frisch 1939. Hahn alone received the Nobel Prize in Chemistry in 1944 for his discovery of the fission of heavy nuclei. Frisch joined the Manhattan Project in the United States in 1943. Meitner refused to participate in the development of atomic weapons and was unfairly written out of the history of atomic energy. She did receive some belated recognition in 1966, two years before her death, when she shared the Enrico Fermi Award. Sime 1996.

14. Anderson, Fermi, and Szilard 1939.

15. Szilard's guess was right, but Nazi Germany abandoned the project before the end of World War II. Much has been written about the reasons for this, but the Germans probably believed that they could win the war using more

conventional weapons without having to invest in the expensive, lengthy, and risky development of an atomic bomb.

16. For book-length accounts of the Manhattan Project, see, e.g., Serber 1998; Bird and Sherwin 2006.

17. For book-length histories of molecular biology, see, e.g., Judson 1979; Kay 1993.

4. ATOMS IN BIOLOGY

1. Max Delbrück interview, 1978, Oral History Project, California Institute of Technology Archives (http://resolver.caltech.edu/CaltechOH:OH_Delbruck_M).

2. Timoféeff, Zimmer, and Delbrück 1935. Amazingly, that paper—and Delbrück—received much publicity when Erwin Schrödinger analyzed it in detail almost ten years later in his famous 1944 book *What Is Life?* That paper, written in German, has recently been translated into English and analyzed further. Sloan and Fogel 2011.

3. An extraordinary character, Félix d'Herelle was a French-Canadian microbiologist. See Summers 1999.

4. Luria 1984.

5. Franco Rasetti had worked with Lise Meitner at the KWI in Berlin-Dahlem from 1931 to 1932 and again from 1933 to 1934. Delbrück was an assistant to Meitner from 1932 to 1937. Rasetti must have been aware of Delbrück's interest in gene mutations and received a reprint of the 1935 paper by Timoféeff, Zimmer, and Delbrück. In his 1982 oral history Rasetti never mentioned meeting Delbrück in Meitner's lab or ever having met Luria in Rome. Rasetti Franco interview, 1982, Oral History Project, California Institute of Technology Archives (http://resolver.caltech.edu/CaltechOH:OH_Rasetti_F).

6. Franco Rasetti, who took the Fermis to the train station, recalled their departure for Stockholm and shared their secret that they would not be returning to Italy.

7. A collection of Cold Spring Harbor Symposia red books between 1941 and 1966 recounts the history of molecular genetics. See http://library.cshl.edu/symposia.

8. Thousands of U.S. residents who were citizens of Germany, Italy, and especially Japan were interned in U.S. camps during World War II.

9. Luria and Delbrück 1943.

10. Mark Adams also taught a phage course at New York University College of Medicine and wrote the detailed "Methods of Study of Bacterial Viruses" in 1950. He authored the classic book *Bacteriophages*, which was published after his premature death in 1956, at age forty-four. His book was completed and edited by his friends and colleagues, with A. Hershey as the principal editor. Adams 1959.

11. Phage research was very active in the 1920s but was essentially abandoned by the mid-1930s. The renewed popularity of phage research after World War II was largely due to the Delbrück's phage course at Cold Spring Harbor.

12. Feld and Szilard 1972, xix.

13. Feld and Szilard 1972, xi.

14. Puck to Szilard, December 12, 1955, LSP b15/f25.

15. Woodward and Doering 1944.

16. The claim that total synthesis was achieved was controversial for many years, but it was fully validated in 2008 in an article dedicated to Doering on his ninetieth birthday. Ball 2008.

17. In an email to the author dated March 11, 2010, Doering clarified how his friendship with Szilard began, and how that friendship resulted in the Doering-Szilard memorandum. SBP.

18. This memorandum is published in Feld and Szilard 1972. However, the confidential appendix attached to the memorandum was heavily edited for publication, with all paragraphs that could identify the people proposed as members removed.

19. Canfield to Cowles Sr., November 19, 1956, LSP b6/f4.

20. Such frenetic networking was typical of Szilard's method of operating. This particular group, however, later became unusually close-knit. Linked by their professional, political, and social interests, the Cowles, Canfields, and Doering would establish family ties as well. The Cowles had two children, John Jr. and Sarah. Some years after these events, Sarah married Doering and John Jr. married Canfield's stepdaughter, Sage Fuller. Doering would remain a friend of Szilard and join him in other undertakings, including the Council for a Livable World, a lobby founded in 1962 to control the nuclear arms race by influencing congressional elections.

21. Such an institute, combining basic and applied research under the same roof, was never realized at the Salk Institute. It would have required a very large and private endowment, which the Salk never had. Such institutions are rare even today, but a good example is the Ludwig Institute for Cancer Research, which focuses on translating cancer research into possible cures. A private source of funding was essential because, until 1980, inventions arising through research supported by U.S. government funds were owned by the government. In December 1980, the U.S. Congress passed the Bayh-Dole Act, which allowed universities and nonprofit research institutions to retain ownership of inventions funded by federal government grants. In return the institutions became responsible for obtaining patents, sharing royalties with the inventors/scientists, and committing the remainder for funding research. Many such inventions are licensed to existing companies, but today it is also common for scientists doing research funded by taxpayer's money to become entrepreneurs, seeking venture capital and founding biotech companies that in some cases become lucrative.

22. Szilard to Salk, January 14, 1957, LSP b17/f6.

23. Lanouette 1992, 399.

24. Salk to Canfield (cc: Szilard), February 8, 1957, MODA, BOC Series 12 Box 6-JS 1957.

25. Lanouette (1992, 400) mentions that in the fall of 1957 Szilard met with Benzer, Brenner, Dulbecco, and Cohn to discuss a new institute to be headed by Salk (statement based on a Brenner interview, p. 548, n. 82). None of those four scientists was in contact with Salk at that time.

26. Szilard to Canfield et al., October 2, 1957, LSP b6/f4.

27. Szilard to Canfield et al., April 20, 1958, LSP b71/f14.

28. On November 1, 1960, at the request of Szilard, Cass Canfield's secretary mailed to Melvin Cohn an intact, unedited version of that appendix. Unlike the edited version published in Feld and Szilard 1972, this document reveals all the names of the scientists considered for membership of the proposed research institute. MCP.

29. It is possible that the memo was also sent to others, perhaps Dulbecco or Lennox. Their papers, however, are not available. The memo preserved in the JSP and LSP is the edited version.

30. Lanouette 1992, 176.

31. J. Monod, foreword to Feld and Szilard 1972.

5. WHAT WAS IT ABOUT LA JOLLA?

1. These examples of misspellings of "La Jolla" are found in archives dating from the late 1950s and early 1960s.

2. Brown to Salk, July 23, 1956, and October 12, 1956, JSP b171/f8.

3. Lorraine Friedman's 1957 diary with notes about her first visit to La Jolla and photos of Salk's 1959 visit were found in Friedman's La Jolla home after her death. They were kindly communicated to the author by her nephew, Jim Friedman. Lorraine's diary entry in July 1957 does not specify the station at which she arrived, only that she called Lenora from the station. Since the San Diego station is quite large, it is likely that she got off the train at the tiny Del Mar station, where most visitors to La Jolla descended in those days.

4. The identity of O'Connor's informant was established with the help of David Rose, the archivist at the March of Dimes.

5. Ohmer to O'Connor, May 1, 1959, JSP b349/f7.

6. The Hoover Pavilion was built as part of the Palo Alto Hospital in the 1930s. It was eventually renovated for use as a medical office and clinic building, although its historic character and art deco exterior style were preserved. See www.cityofpaloalto.org.

7. O'Connor to Sterling, July 21, 1959, JSP b350/f1.

8. Cohn's recollections about Salk's visit are based on several conversations with the author in 2009 and 2010.

9. Szilard to Salk, May 8, 1959, and Memo, May 7, 1959, JSP b349/f7.

10. In the 1950s Pittsburgh was so polluted that it deserved its nickname "Hell with the Lid Off." Today, however, the city with three rivers is a shining example of the success of the Clean Air Act and Clean Water Act.

11. Revelle and Szilard were friends who shared many social and political concerns. Like Szilard, Revelle was deeply involved with the population problem as it relates to the conservation of resources and the protection of the environment, and both were participants in the Pugwash Conferences on Science and World Affairs. Revelle eventually became director of the Center for Population Studies at Harvard.

12. Salk note to Friedman, June 26, 1959, JSP b344/f1.

13. Salk statement for Sterling, June 16, 1959, JSP b350/f4.

14. Salk to Murrow, August 6, 1959, JSP b350/f4.

15. The chronology of Salk's early visits to La Jolla has been preserved in the archives of the UCSD Office of the Chancellor. Revelle to Wellman, April 3, 1960, CHAN b227/f5.

16. Originally built as a summer house by Ellen's parents in 1922, this historic home was greatly expanded by the Revelles. Roger died in 1991 and Ellen in 2009. In July 2010 their house was put on the market for the first time at a price of fourteen million dollars.

17. Revelle to Kerr, August 23, 1959, JSP b350/f1.

18. Crane 1991.

19. The shells used for artillery training at the time were filled with sand instead of explosives.

20. After the war there was a severe shortage of lumber. The buildings of Camp Callan were purchased by the city and sold at a low price to provide building materials for the homes of servicemen who decided to settle in San Diego.

21. De Hoffmann became president of General Atomic in 1958 and president of the Salk Institute in 1972. He is discussed further in chapter 12.

22. TRIGA stood for Training, Research, Isotopes, General Atomic.

23. Freeman Dyson has many fond memories of the year he spent at General Atomic in its early days. Dyson 1981.

24. The General Atomic building was built by the Los Angeles architects Pereira & Luckman.

25. Anderson 1993.

26. Raitt and Moulton 1967; Shor 1978.

27. Anderson 1993.

28. Salk handwritten notes, JSP b344/f1.

29. Salk to Simnad, September 25, 1959, JSP b349/f7.

30. This list was reconstructed on the basis of various correspondence in JSP. Watson also reports meeting Meselson and Cohn in La Jolla in early March 1960. Watson 2007, 145.

31. Salk to Murrow, January 20, 1960, JSP b350/f4.

32. Several pictures of that visit were found in the house of Lorraine Friedman and communicated to the author by her nephew Jim Friedman. One of those photos was published later, in the January 21, 1960, issue of the *La Jolla Light* (see fig. 1).

33. Revelle to Wellman, April 3, 1960, CHAN b227/f5.

34. Salk's statement, March 15, 1960, JSP b70/f9.

35. Anecdote reported by Jim Friedman to the author in March 2007.

36. O'Connor's statement, March 15, 1960, JSP b70/f9.

37. Ellen Revelle Eckis interview, August 10, 1998, SIOA, SIO Reference Series 99-12.

38. Gallison to Salk, May 25, 1960, JSP b350/f2.

39. Whitney to Salk, June 9, 1960, JSP b350/f2.

40. Today the road is called Torrey Pines Scenic Drive. The north half of the parcel is assigned to the university. This is the site of the new Stem Cell Building, which is shared by the four institutions that formed the Sanford Consortium for Regenerative Medicine: the Sanford-Burnham Institute, the Salk Institute, the Scripps Research Institute, and UCSD.

41. Puck to Salk, April 16, 1960, JSP b404/f1.
42. Melvin Cohn, conversations with the author.
43. MCP.

6. THE PASTEUR CONNECTION

1. Lawrence 1999.
2. Pappenheimer 1937.
3. This led to the famous Avery-MacLeod-McCarty experiment published in 1944, which demonstrated that the genetic material is DNA. Avery, MacLeod, and McCarty 1944.
4. For detailed biographical information about Monod, see Lwoff and Ullmann 1979; Debré 1996; Soulier 1997.
5. Philippe Monod studied law at the University of Aix-en-Provence. He later specialized in international law at Harvard and became an ambassador.
6. This saved Monod's life, as the *Pourquoi-Pas?* sunk in a terrible storm off the coast of Iceland.
7. De Lattre de Tassigny 1971, 654.
8. The Clairol operations in Connecticut were called Lawrence Richard Bruce Inc. in the early 1940s.
9. Information about Cohn's World War II battles, campaigns, and service outside the United States is based on his Army Separation Records. MCP.
10. Japan did not respect the Geneva Conventions protecting medical personnel.
11. Cohn, conversation with the author, ca. 2009.
12. Years later Kengo became a key research associate in Cohn's laboratory, first at Washington University and later at the Salk Institute.
13. MCP.
14. The GI Bill of Rights was signed into law by FDR in June 1944.
15. Baldwin and Ferry 1994.
16. For an overview of Monod's early work, see Lwoff and Ullmann 1978.
17. For accounts of his time at the Pasteur institute, see Cohn, 1979, 1989; Judson 1979.
18. Hogness, Cohn, and Monod 1955.
19. Cohn and Monod 1951.
20. Judson 1979; Jacob 1988.
21. Germaine Bazire eventually married Roger Stanier and Anna-Maria Torriani married Luigi Gorini.
22. Jacob 1988, 216.
23. Szilard did not meet Cohn in Paris because his first trip to Europe since leaving in 1938 was in the fall of 1957, after Cohn had left Paris. Gertrud Szilard to Monod, February 13, 1968, JMP Cor. 16.
24. Lennox, interview by the author, June 2008, SBP.
25. For a detailed account of Los Alamos life and events, see Serber 1998.
26. Glauber won the Nobel Prize in 2005.
27. Bird and Sherwin 2006.
28. Lennox, Luria, and Benzer 1954; Lennox 1955.

29. This may have been the summer of 1955 or 1956.

30. Their first joint paper was Attardi, Cohn, Horibata, and Lennox 1959.

31. Lennox and Cohn kept in touch and Lennox visited Cohn in January 1960 for the Antibody Workshop at Stanford (see chapter 9). Lennox, interview by the author, June 2008, and Antibody Workshop archives, MCP and SBP.

32. Monod to Pappenheimer, January 30, 1959, MCP.

33. Williams to Cohn, February 2, 1960; Pappenheimer to Cohn, March 10, 1960, MCP.

34. Pappenheimer to Cohn, May 11, 1960, MCP.

35. Stanier to Cohn, June 9, 1960, MCP.

36. Monod and Cohn exchanged an extensive correspondence starting in 1949. The original letters are preserved in the Pasteur Institute Archives, JMP, and MCP.

37. Cohn to Monod, March 1960; Monod to Cohn, April 1, 1960, JMP and MCP.

38. Cohn to Monod, July 6, 1960, JMP and MCP.

39. Monod to Cohn, August 8, 1960, JMP and MCP.

40. NSF grant application, 1961, JMP Ser. 06.

41. Monod to Cohn, November 19, 1958, JMP and MCP.

42. Brunerie, *récit autobiographique,* entries for January 18, 1965 (p. 180) and February 7, 1966 (p. 222), Pasteur Institute Archives, fonds Brunerie, BRU 1.

43. Monod to Cohn, June 3, 1955, JMP and MCP.

44. Crick 1979.

7. THE SPIRIT OF PARIS

1. Crick 1988, 145.

2. Watson to Cohn, November 23, 1960, MCP.

3. Warren Weaver was not related to Harry M. Weaver, the research director of the National Foundation for Infantile Paralysis (see chapter 1).

4. Weaver coined the term "molecular biology" in his 1938 annual report to the Rockefeller Foundation. The term was reinvented and propagated by W. T. Astbury and by P. A. Weiss in the early 1950s. See Weaver 1970a; Astbury 1961; Weiss 1971, 270.

5. Weaver 1970b.

6. Mason and Weaver 1929.

7. "We will be able to buy a lot of things with the Rockefeller money, and my fingers are itching when I think of the beautiful experiments we will be able to do with the new equipment." J. Monod to O. Monod, July 24, 1946, JMP, cited in Gaudilliere 2002, 31.

8. Weaver 1970b, 90.

9. Rees 1980.

10. The article appeared in the July and October 1948 issues.

11. Shannon and Weaver 1949.

12. Weaver 1964.

13. Weaver 1964, 12.

14. Weaver 1970b, 128.

15. Weaver to Salk, April 6, 1960, JSP b350/f1.

16. Salk to Weaver, April 13, 1960, JSP b350/f1.

17. Salk to Weaver, April 11, 1960, drafts for April 13 letter. These drafts include suggestions from Leo Szilard. JSP b350/f1.

18. "Research Institute to Be Established in California" 1960.

19. "Announcements" 1962.

20. Crick, Watson, and Wilkins shared the Nobel Prize in Physiology or Medicine in 1962.

21. For a book-length biography of Crick, see Olby 2009.

22. Watson and Crick 1953a and 1953b.

23. Salk to Buzzati-Traverso, June 6, 1960, JSP b350/f2.

24. Shor 1978, 212.

25. For the history of Buzzati-Traverso's International Laboratory of Genetics and Biophysics in Naples, see Levi-Montalcini 1988, 193; Capocci and Corbellini 2002.

26. One might expect that Weaver and Bronowski met in London during World War II, but there is no mention of any early acquaintance in Weaver's autobiography.

27. Weaver 1970b, 130.

28. Some of his early books were *The Poet's Defense* (1939), *William Blake: A Man Without a Mask* (1943), *The Common Sense of Science* (1951), and *The Face of Violence* (1954).

29. First published in 1956 in *University Quarterly* and *The Nation*.

30. Bronowski 1959.

31. Snow 1959.

32. Accounts of Salk's meetings in Europe in July 1960 are based on transcripts of notes dictated during his stay in London and communicated to the author by Jim Friedman. SBP.

33. Interview of Rita Bronowski in Orrick 1996.

34. See the history of General Atomic in chapter 5.

35. The 1959 edition of *Science and Human Values* and the cover letter from Lorraine were found in the Cohn papers. L. Friedman to M. Cohn, August 17, 1960, MCP. Several of the people involved received the same. JSP b404/f1.

36. Judson 1979; Alexander 1992; Morange 1998; Kay 2000.

37. Brunerie, *récit autobiographique,* p. 96. Pasteur Institute Archives, fonds Brunerie, BRU 1.

38. Salk to Monod, October 17, 1961, ESIA D4/41.

39. Salk to Snow, Salk to Monod, and Salk to Crick, October 17, 1961, ESIA D4/41.

40. Flexner 1960. See chapter 2.

41. Bronowski to Salk, December 12, 1961, JSP b396/f8.

42. Salk to Monod, October 23, 1961, JMP Mon. Ins. 05, 1961.

43. JSP b283/f4.

44. Olby 2009, 366; Lanouette 1992, 474.

45. Crick 1979.

46. Salk to Monod, December 8, 1961, JMP Mon. Ins. 05, 1961.

47. Minutes attached to Bronowski to Monod, February 20, 1962, JMP Mon. Ins. 05, 1962.

48. Crick to Monod, May 18, 1962, JMP Mon. Ins. 05, 1962.

49. ESIA D12/5.

50. "Announcements" 1962.

8. OUR DEAR KAHN BUILDING

1. From Churchill's October 1943 address to the House of Lords to consider how to rebuild the old House of Commons, which had been destroyed by German bombs in May 1941. It was eventually rebuilt in 1950 in its old form, even though it had an insufficient number of seats for all its members!

2. "Polio Fund Owes $2,000,000" 1960.

3. Interview with Charles L. Massey on October 3, 1983, MODA Oral History Records.

4. First Annual Report of the Director, March 7, 1962, ESIA D19/f13.

5. Interview of Jonas Salk on September 8, 1984, MODA Oral History Records.

6. Jack MacAllister, interview by the author, April 2010, SBP. See also Leslie 2005; Wiseman 2007.

7. First Annual Report of the Director, March 7, 1962, ESIA D19/f13.

8. See, e.g., Leslie 2005, 134.

9. Dictated notes transcribed by Lorraine Friedman and communicated to the author by Jim Friedman, SBP.

10. Salk to Puck, February 28, 1961, JSP b404/f2.

11. Annual Report of the Director, February 1971, SBP.

12. ESIA D10/f3.

13. MODA Series 1: Correspondence: O' Connor to NF Board of Trustees, December 8, 1961.

14. MODA Series 1: Correspondence: Planning and Development: News Conference, May 31, 1962.

15. "Announcements" 1962.

16. This explains why Piel took over the presidency from Salk, who, as president, would have had to raise money for an institute bearing his own name.

17. Puck had accepted an appointment as a Fellow but requested that his name not be used in publicity. He never joined the Salk Institute and eventually became director of the University of Colorado Cancer Center in Denver.

18. Benzer had accepted an appointment as a Fellow, he participated in the press conference, and he wanted his name mentioned in publicity. However, he never joined the Salk Institute and instead took a position at Caltech.

19. MCP. The rabbi's speech inspired his wife, Sally Cohn, to organize an association of prominent local women, including the wife of the mayor, Mrs. Charles Dail, and of the chancellor of UCSD, Mrs. Herbert York. This Women's Association for the Salk Institute (later the Salk Institute Association) supported the institute for over four decades, raising about one million dollars and contributing countless volunteer hours.

20. Discussed in Oshinsky 2005.

21. An improved Salk vaccine has been used in the United States since 1999. See chapter 1.

22. Weaver to Heald, July 18, 1963; Salk to Heald, September 20, 1963, JSP b395/f8.

23. Interview with Weaver, February 27, 1969, JBP.

24. Minutes of the Fellows Meeting, March 30–31, 1963, ESIA D12/f5.

25. Bonner to Salk, Cohn, Dulbecco, and Lennox, March 11, 1963, ESIA D12/f5.

26. Lennox to Szilard, March 24, 1963, LSP b11/f31.

27. Annual Report of the Director, February 1971, SBP.

28. First Annual Report of the Director, March 7, 1962, p. 10, ESIA D19/f13.

29. Wiseman 2007.

30. Numerous books and articles have been written by architects describing the features and construction of the Salk Institute building and the partnership of Salk and Kahn. See, e.g., Friedman 1991; Leslie 2005; Wiseman 2007.

31. Today that road is known as North Torrey Pines Road.

32. Kahn proposed this arrangement at the San Diego city hearings on March 15, 1960, before a precise allocation of the site had been agreed upon. Kahn must have been attracted by the topography of that parcel divided by a canyon. The agreement with the city on the splitting of the site between the Salk Institute and UCSD, however, did not take place until May 24, 1960. See chapter 5 and Johns 1960.

33. The architectural contract between Kahn and the institute, signed in July 1961, was amended in August 1963 to stop work on the meetinghouse and residences. Today the south extension site where residences were planned is considered a conservation easement, not to be built upon. The north extension, however, could still be built, and, although plans have been halted for now, a meetinghouse is shown on that location in the latest master plan.

34. Larson 2000.

35. Earl Walls, interview by the author, April 2011, SBP.

36. In the estimate of fifteen million dollars, twelve million had been budgeted for the laboratory building and three million for the meetinghouse and residences.

37. The NF used funds collected through the mechanism of the MOD and eventually repaid this loan plus interest over ten years.

38. See chapter 9. As is common with functional temporary structures, the barracks built in 1963 are still standing as of this writing (2011).

39. Numerous books have narrated the role of the Mexican architect Luis Barragán in replacing the planned landscaped garden with an empty plaza. See, e.g., Leslie 2005, 165. While Barragán's visit in February 1966 has passed into legend, my diary confirms that Salk, Kahn, MacAllister, Cohn, and I took Barragán to dinner at La Favorite, the French restaurant in La Jolla, on February 24, 1966. SBP.

40. Leslie 2005, 168.

41. "It looks like a prison."

42. The La Jolla Historical Society (www.lajollahistory.org) has produced a fine DVD on this subject, *The Story of Historic Torrey Pines Gliderport.*

43. See www.magritte.be.

44. Roger Guillemin, for example, was mesmerized by the extraordinary beauty of the Salk Institute (Guillemin 1998). Gildred was the first donor who was seduced by the Kahn building.

45. Interview with Weaver, February 27, 1969, JBP.

9. PIONEERING

1. Crick to Salk, November 25, 1963, ESIA D4/f41.

2. Crick 1988.

3. Hubel 1963.

4. Hubel and Wiesel 1962.

5. Crick 1988.

6. Dulbecco 1990; Kevles 1993; Renato Dulbecco interview, 1998, Oral History Project, California Institute of Technology Archives.

7. Dulbecco 1990, 107; Renato Dulbecco interview, 1998, Oral History Project, California Institute of Technology Archives, p. 9.

8. Levi-Montalcini 1988.

9. Levi-Montalcini 1988, 113.

10. By chance Rita Levi-Montalcini was on the same ship, heading for St. Louis on her first visit to Washington University. In her autobiography (1988, 117) she mentions that this trip took place in 1946, but this must be an error, since the *Sobieski* served as an Allied troopship in World War II and did not resume commercial service until May 1947. See www.theshipslist.com/ships/descriptions.

11. "A very smart student, but one that had a somewhat strange behavior." Dulbecco 1990, 138.

12. The role of Delbrück in getting Dulbecco involved in animal virus work and the details of Dulbecco's trip have been described by Kevles (1993) on the basis of the Delbrück papers.

13. Dulbecco 1952.

14. Dulbecco and Vogt 1954.

15. ESIA D12/f5.

16. The Harvard Neurobiology Group spent the summers of 1967 to 1971 at the Salk Institute giving seminars and educating our faculty and trainees in the new field of neurobiology.

17. Bourgeois, Cohn, and Orgel 1966.

18. Szilard–Salk correspondence 1957–1963, LSP b17/f6.

19. The unfinished Kahn courtyard reminded me of my school in ruins after a bombardment of Brussels.

20. The original version was eventually published in the German edition of *The Voice of the Dolphins* twelve years after Leo's death. Szilard 1976.

21. Weart and Szilard 1978.

22. Szilard to Weiss-Szilard, August 4, 1960, LSP b114/f26.

23. Szilard 1959; Orgel 1963.

24. ESIA D4/f36.

25. Adloff and Kauffman 2008.

26. In England some two million children were evacuated from big cities to the countryside during World War II. Unfortunately, massive evacuation could not be organized in Belgium because it was occupied by Nazi Germany.

27. Watson and Crick 1953a and 1953b.

28. Orgel, interview by the author, April 2006, SBP.

29. Krohn 1966.

30. Hayflick 1966.

31. Telomeres are structures that protect the ends of chromosomes. A Nobel Prize was awarded in 2009 for discoveries of mechanisms by which telomeres protect chromosomes and are synthesized. One of the current (2011) nonresident Fellows of the Salk Institute, Elizabeth Blackburn, shared that award with Carol Greider and Jack Szostak.

32. See, for example, Doolittle 1974; Porter 1986; Podolsky and Tauber 1997; Potter 2007.

33. Their first joint paper was Attardi, Cohn, Horibata, and Lennox 1959.

34. Herman Eisen, Fred Karush, Rodney Porter, and Jon Singer were coinvestigators on that grant and participated in the organization of the workshops. Archives of the early Antibody Workshops and the NSF grant application are kept in the MCP.

35. The complete list of invitees and participants as well as the program are in MCP as well as in the author's possession. SBP.

36. According to the author's diary, it was on Sunday, November 8, 1964. SBP.

37. In the mid-1800s Warner's Ranch was a stop on the twenty-five-day Butterfield stagecoach run from Missouri to San Francisco. See http://parks.ca. gov/?page_id=25444.

38. The Cohn letter of invitation, dated December 8, 1964, and the answers of the invitees are in MCP.

39. The 1964 color pamphlet for Warner Springs is preserved in the author's papers. SBP.

40. Cohn first heard about the work on immunoglobulin sequences being done in Craig's lab in a telephone conversation with Stanford Moore and received confirmation in a letter from Rodney Porter. Porter to Cohn, October 20, 1964, MCP.

41. Porter to Cohn, October 20, 1964, MCP.

42. Actually it was a chain from a human Bence-Jones protein corresponding to the light chain of an antibody.

43. Doolittle 1974.

44. Hilschmann and Craig 1965.

45. Ochoa shared the Nobel Prize in Medicine with Arthur Kornberg in 1959, and Khorana shared it with Robert Holley and Marshall Nirenberg in 1968.

46. Neurophysiology Group—Correspondence, ESIA D11/f11.

47. In 1966 Harvard Medical School made a counteroffer and established a Department of Neurobiology with Kuffler as chair.

48. Annual Fellows Meetings 1962–1965, ESIA D12/f5; Annual Trustees Meeting 1965, ESIA D10/f10.

49. September 28, 1965, Trustees Meeting, ESIA D10/f12.

50. In addition two trustees, Joseph Nee and Jerome Hardy, served very briefly as CEO and president, respectively, in 1967.

51. See the colloquium program and participants in JSP b360/f12.

52. JSP b403/f12.

53. Eventually an institute employee pleaded guilty and did time in a federal prison.

54. In David Baltimore's biography, Crotty (2001) mentions the art show incident and David's state of mind while at the Salk Institute in 1965–67. Today Baltimore is a nonresident Fellow of the institute.

55. JSP b359/f5.

56. Orgel to Salk, May 3, 1967, JSP b359/f5.

57. Stroll to Kinzel, May 3, 1967, JSP b359/f5.

58. Salk's handwritten draft specifies that this text was not used. JSP b359/f5.

10. THE MCCLOY BOYS

1. This quote from Pope was used by Jonas Salk to describe his high hopes for Slater's presidency after the painful end of Kinzel's.

2. Slater's biographical information is based on Hyman 1975 and on the author's personal recollections.

3. Bird (1992) wrote a book-length biography of McCloy.

4. Berghahn 2001 discusses the role of McCloy and his team in postwar Europe.

5. Bird 1992, 440.

6. Hyman 1975, 234–35.

7. John Hunt, interview by the author, July 29, 2006, SBP.

8. Hyman 1975. See also www.aspeninstitute.org/.

9. Robert Hutchins appeared in chapter 4 as a friend and supporter of Leo Szilard.

10. Hyman 1975, 239. It is surprising that Slater would give a lecture about biology since he had no credentials as a biologist or a scientist. It should be mentioned that Hyman's book was published in 1975 in honor of the Aspen Institute's twenty-fifth anniversary while Slater was its president. Hyman's book does not cite sources such as archives, so it seems that Hyman's facts concerning Slater are based entirely on Slater's own statements.

11. Memorandum dated September 11, 1967, ESIA D9/f44.

12. Hardy to Slater, October 26, 1967, ESIA D6/f30.

13. The original note is preserved in the author's papers, SBP.

14. See chapter 9 and JSP b403/f12.

15. 1968 Annual Fellows Meeting, ESIA D12/f13.

16. Hunt 1956, 1968.

17. The fascinating history of the CCF after World War II has been told in a number of books and articles. See, for example, Berghahn 2001; Bird 1992; Coleman 1989; Grémion 1995; Saunders 1999; Shils 1990. A discussion of the CCF is outside the scope of this book, but several important characters in the story of the Salk Institute were a product of that era of the intellectual cold war.

18. During this tense post–World War II period the U.S. government used private foundations to channel American financial support into projects of political importance. Bird 1992, 426–29; John Hunt, interview by the author, July 29 and 31, 2006, SBP.

19. This was on February 15, 1968. Author's diary and John Hunt, interview by the author, July 31, 2006, SBP.

20. ESIA D6/f37.

21. Luria to Hunt, May 30, 1969, with a copy and a note to Monod, JMP.

22. ESIA D13/f19–f22.

23. In 1970 the Council for Biology in Human Affairs was created to coordinate the Salk Institute's new programs. See chapter 11.

24. Lehrman was a key participant in the February 1967 Salk Institute colloquium Studies on Development of Behavior. See chapter 9.

25. For work done in 1965 at Cornell University, Robert W. Holley shared the 1968 Nobel Prize in Medicine with Gobind Khorana and Marshall W. Nirenberg. Holley was to found the NIH Cancer Research Center at the Salk Institute in 1973.

26. According to Crotty 2001, 69–70, the fact that junior scientists had no say in the governance of the Salk Institute in 1967 was a major source of resentment for David Baltimore at the time (see chapter 9). Actually, junior faculty positions did not yet exist in 1967, as the institute was still in its infancy. The first five junior Members were appointed by the board in February 1970 (ESIA D10/f30). They were Ursula Bellugi, Suzanne Bourgeois, Walter Eckhart, Stephen Heinemann, and David Schubert. All five original Members are still active at the Salk Institute today (2012).

27. ESIA D6/f47.

28. The invitation, program, cover letter, and information from Dr. Lee are preserved in the author's papers together with notes taken at the meeting. SBP.

29. For autobiographical essays see Guillemin 1978 and 1998.

30. Author's diary, January 14, 1969, SBP.

31. Minutes of the Annual Fellows Meeting of February 8–9, 1969, ESIA D12/f15.

32. Guillemin, interviews by the author in April 2006, SBP.

33. ESIA D10/f29.

34. Schubert, interviews by the author, June, July, and October 2009, SBP.

35. ESIA D9/f44.

36. ESIA D10/f29.

37. ESIA D10/f30.

38. Jonas and his wife, Donna, had drifted apart and were divorced in 1968.

39. Chantal Hunt, interview by the author, August 1, 2006; Françoise Gilot, interview by the author, August 9, 2009, SBP.

40. Chantal Hunt's two children were from a first marriage, and Françoise Gilot's two children, Paloma and Claude, were fathered by Pablo Picasso. In 1955 Gilot married a young painter of her generation, Luc Simon, with whom she had a second daughter, Aurelia Simon, in 1956. Aurelia was not born at the time Françoise Gilot and Chantal Hunt met. Chantal Hunt, interview by the author, August 2006, SBP.

41. Gilot and Lake 1964.

42. Françoise Gilot, interview by the author, August 9, 2009, SBP.

43. ESIA D11/f42.

44. ESIA D10/f35.

45. The cause of death was "heart ailment complicated by pneumonia" according to *National Foundation News* 1972.

46. Glasser to McCloy, May 10, 1972; McCloy to Glasser, May 25, 1972, MODA. One of the conditions was that Jonas Salk carry the title of founding director of the institute while relinquishing his previous duties and responsibilities as director to the president and CEO.

47. Dulbecco shared the Nobel Prize with David Baltimore and Howard Temin in 1975, and he returned to the Salk Institute in 1977.

48. John Hunt's insights were very helpful in interpreting Slater's motivation. John Hunt, interview by the author, 2006, SBP.

49. Slater remained president of the Aspen Institute for seventeen years, during which he transformed it into an international network of institutes. Not surprisingly, the first Aspen Institute outside the United States was established in Berlin in 1974, and Shepard Stone was its director.

50. John Hunt, interview by the author, July 2006, SBP.

11. BIOLOGY IN HUMAN AFFAIRS

1. "I am a man: I consider nothing that is human to be alien to me." This famous line appeared in a play by Terence, a Roman dramatist from the second century B.C. It is often mistakenly attributed to the French Renaissance writer Montaigne. For a recent commentary, see Strickland 2005.

2. See, for example, Bronowski 1967.

3. Kay mentions that Jakobson visited the institute between 1962 and 1965 and intended to move to the Salk. Those dates, however, are wrong, and Jakobson never considered a full-time position at the institute, although he did consider moving to UCSD. See Kay 2000, 306, and Jakobson-Bronowski correspondence, ESIA D4/f72.

4. MCP.

5. See Monod to Popper, August 23, 1972, JMP Cor. 13; Popper 1973.

6. See Cohn 1967.

7. For the complete catalog of the Szilard Memorial Collection as of the early 1970s, see ESIA D16/f27–30.

8. For a description of the Archive of Contemporary Biology, see ESIA D16/f32. The long-term plan was to set up at the institute an appropriate storage facility to preserve these archives as well as the papers of the Fellows.

9. Only two occasional papers were ever published by the Salk Institute: a translation of Monod's inaugural lecture at the Collège de France entitled "From Biology to Ethics" (1969) and a paper by Hans Gaffron, "Resistance to Knowledge" (1970). At least four other occasional papers were mentioned as being prepared in 1971. See JSP b421/f5. Copies of the Monod and Gaffron papers are preserved in the author's collection, SBP.

10. For more background on the CBHA, see ESIA D11/f42.

11. For Boardman's biography see ESIA D4/f81.

12. See the Council on Foreign Relations website at www.cfr.org.

13. For a summary of CBHA activities, see JSP b421/f5.

14. This was allowed because most of the Fellows were adjunct professors at UCSD.

15. Email correspondence between M. Jacobson and the author, July 2009, SBP.

16. Jacobson 1972.

17. For more on the CSPI, see www.cspinet.org.

18. Crichton 1969. See also www.michaelcrichton.net.

19. ESIA D17/f10.

20. Programs and Activities of the CBHA, JSP b421/f5.

21. Friedmann, interviews (July 2006) and conversations with the author, SBP.

22. Friedmann 1971; Friedmann and Roblin 1972. Both papers acknowledge the authors' affiliation with the CBHA.

23. For the history of the recombinant DNA issue and the involvement of Roblin, see Fredrickson 2001. See also Paul Berg, interview by the author, November 22, 2011, SBP.

24. See http://www-pediatrics.ucsd.edu/Research/Labs/Dr Theodore Friedmann - Gene T/Pages/default.aspx

25. Platt 1969.

26. See Sher, interview by the author, November 12, 2008, and email correspondence between the author, Donato Cioli, and Paul Knopf, July 2009, SBP. See also Sher 1981.

27. Weir 1969.

28. Morin had written a series of articles on the subject for the French newspaper *Le Monde,* and Bronowski had also analyzed the bases of student protests. See Bronowski 1969.

29. JSP b421/f5.

30. Hunt to Monod, May 1, 1969; Monod to Hunt, May 12, 1969, JMP Cor. Ins. 05–10.

31. John Hunt, interview by the author, July 31, 2006, SBP.

32. Morin 1970.

33. Morin, interview by the author, September 12, 2006, SBP. See also the preface to Morin 1973.

34. See www.royaumont.com.

35. Morin and Piattelli-Palmarini 1974.

36. The distinguished achievements of the Centre Royaumont pour une Science de l'Homme are beyond the scope of this book. Their permanent legacy is a collection of scholarly books, some of which have become classic, including Morin and Piattelli-Palmarini 1974; Sullerot 1978; Piattelli-Palmarini 1979.

37. John Hunt, interview by author, July 2006, and Edgar Morin, interview by author, September 2006, SBP.

12. A NAPOLEON FROM BYZANTIUM

1. "We owe it to our Führer." Schrödinger (1992) was referring to the wonderful time he spent in Dublin during World War II.

2. George Dyson (2002) has written the most detailed account of de Hoffmann's early activities, from his time at Los Alamos to his work at General Atomic.

3. De Hoffmann's biographical data is based on records obtained from the city of Vienna and the Vienna Jewish Archives. In the United States, information was obtained from the U.S. Citizenship and Immigration Services and the National Archives.

4. Vagujhely was a largely Jewish settlement that used to be in Slovakia but became part of Hungary during the Austro-Hungarian dual monarchy. Before powerful anti-Semitism reached that region, eastern Jews were welcome. Most were wealthy merchants who received titles and positions as civil servants in exchange for making favorable loans to ruined Magyar aristocrats (see chapter 3). It appears likely that de Hoffmann's grandfather received the title von Vagujhely at that time. Today the former town of Vagujhely is called Nové Mesto nad Váhom and is located in the Slovak Republic.

5. The SS *Nea Hellas* served as a British war transport until 1946.

6. For a history of the Manhattan Project and its people, see the Los Alamos National Laboratory website, www.lanl.gov/history.

7. Teller eventually became unpopular among his colleagues after he testified that Oppenheimer was a security risk. See Bird and Sherwin 2006.

8. Teller 2001, 275.

9. See www.nps.gov/mimi/historyculture/atlas-icbm.htm.

10. See Dyson 2002, 32.

11. Hopkins had died in 1957.

12. Salk to de Hoffmann, January 22, 1960, and July 5, 1960, JSP b350/f2.

13. TRIGA stood for Training, Research and Isotopes, General Atomic. HTGR stood for High Temperature Gas-Cooled Reactor. See Teller 2001, 423; F. Dyson 2004, 140; G. Dyson 2002, 274.

14. G. Dyson 2002.

15. A number of letters from de Hoffmann to Monod are typical of his secretive style. JMP.

EPILOGUE

1. Widely used translation variation of a quote attributed to the Greek philosopher Heraclitus, ca. 500 B.C.

2. As of this writing (2012) the total staff is about one thousand. For current institute programs and faculty, see www.salk.edu.

3. Gruss 2012.

Abbreviations

AMP	Applied Mathematics Panel
AEC	Atomic Energy Commission
AMA	American Medical Association
CBHA	Council for Biology in Human Affairs
CCF	Congress for Cultural Freedom
CCNY	College of the City of New York
CDC	Centers for Disease Control
CSH	Cold Spring Harbor
DNA	Deoxyribonucleic acid
Doc	Nickname of Basil O'Connor
FDA	Food and Drug Administration
FDR	Franklin Delano Roosevelt
GA	General Atomic
GD	General Dynamics
GEB	General Education Board
Gestapo	Geheime Staatspolizei
IACF	International Association for Cultural Freedom
KWI	Kaiser Wilhelm Institute
MOD	March of Dimes
Nazi	Nationalsozialistische Deutsche Arbeiterpartei

NFIP	National Foundation for Infantile Paralysis
NF-MOD	National Foundation—March of Dimes
NIH	National Institutes of Health
NSF	National Science Foundation
NYU	New York University
OSRD	Office of Scientific Research and Development
Pitt	University of Pittsburgh
Polio	Poliomyelitis, Infantile Paralysis
RNA	Ribonucleic acid
SED	Special Engineering Detachment
SIO	Scripps Institute of Oceanography
UC	University of California
UCL	University College of London
UCLA	University of California Los Angeles
UCSD	University of California San Diego
WHO	World Health Organization
WPA	Works Progress Administration

References

ARCHIVES AND COLLECTIONS

CHAN   Office of the Chancellor Archives, UCSD Mandeville Special Collections, San Diego

ESIA   Early Salk Institute Archives, The Salk Institute, San Diego

JBP   Jacob Bronowski Papers, University of Toronto, Toronto

JMP   Jacques Monod Papers, Pasteur Institute Archives, Paris

JSP   Jonas Salk Papers, UCSD Mandeville Special Collections, San Diego

LSP   Leo Szilard Papers, UCSD Mandeville Special Collections, San Diego

MCP   Melvin Cohn Papers, Pasteur Institute Archives and private collection

MODA   March of Dimes Archives, March of Dimes, White Plains, New York

SBP   Suzanne Bourgeois Papers, private collection

SIOA   Scripps Institute of Oceanography Archives, SIO, San Diego

PUBLISHED WORKS

Adams, M. H. 1959. *Bacteriophages*. New York: Interscience Publishers.

Adloff, J.-P., and G. B. Kauffman. 2008. "Leslie Orgel (1927–2007), Pioneer Researcher on the Origin of Life." *Chem Educator* 13: 248–56.

Alberts, R. C. 1986. *Pitt: The Story of the University of Pittsburgh, 1787–1987*. Pittsburgh, PA: University of Pittsburgh Press.

Alexander, Henri. "Jean Brachet and His School." *International Journal of Developmental Biology* 36 (1992): 29–41.

Anderson, H.L., E. Fermi, and L. Szilard. 1939. "Neutron Production and Absorption in Uranium." *Physical Review* 56: 284–86.

Anderson, N.S. 1993. *An Improbable Venture: A History of the University of California, San Diego.* La Jolla, CA: UCSD Press.

"Announcements." 1962. *Science* 136: 769.

Astbury, W.T. 1961. "Molecular Biology or Ultrastructural Biology?" *Nature* 190: 1124.

Attardi, G., M. Cohn, K. Horibata, and E.S. Lennox. 1959. "On the Analysis of Antibody Synthesis at the Cellular Level." *Bacteriological Reviews* 23: 213–23.

Avery, O.T., C.M. MacLeod, and M. McCarty. 1944. "Studies on the Chemical Nature of the Substance Inducing Transformation of Pneumococcal Types. Induction of Transformation by a Desoxyribonucleic Acid Fraction Isolated from Pneumococcus Type III." *Journal of Experimental Medicine* 79: 137–58.

Baldwin, R.L., and J.D. Ferry. 1994. *John Warren Williams 1898–1988.* National Academy of Sciences, Washington, D.C.

Ball, P. 2008. "Quinine Steps Back in Time." *Nature* 451: 1065–66.

Batterson, S. 2006. *Pursuit of Genius: Flexner, Einstein, and the Early Faculty at the Institute for Advanced Study.* Wellesley, MA: A K Peters.

Benison, S., ed. 1967. *Tom Rivers: Reflections on a Life in Medicine and Science; An Oral History Memoir.* Cambridge, MA: MIT Press.

Berghahn, V.R. 2001. *America and the Intellectual Cold Wars in Europe.* Princeton, NJ: Princeton University Press.

Bird, K. 1992. *The Chairman: John J. McCloy: The Making of the American Establishment.* New York: Simon & Schuster.

Bird, K., and M.J. Sherwin. 2006. *American Prometheus: The Triumph and Tragedy of J. Robert Oppenheimer.* New York: Alfred A. Knopf.

Bourgeois, S., M. Cohn, and L.E. Orgel. 1966. "Suppression of and Complementation among Mutants of the Regulatory Gene of the Lactose Operon of Escherichia coli." *Journal of Molecular Biology* 14: 300–302.

Brokaw, T. 1998. *The Greatest Generation.* New York: Random House.

Bronowski, J. 1939. *The Poet's Defence.* Oxford: Oxford University Press.

———. 1943. *William Blake: A Man Without a Mask.* London: Secker & Warburg.

———. 1951. *The Common Sense of Science.* London: Heinemann.

———. 1954. *The Face of Violence.* London: Turnstile Press.

———. 1959. *Science and Human Values: The Impact of Science on Ethics and Human Values as Well as on Our Physical Environment.* New York: Harper & Row.

———. 1967. "Human and Animal Languages." In *To Honor Roman Jakobson: Essays on the Occasion of His Seventieth Birthday,* pp. 374–94. The Hague: Mouton & Co.

———. 1969. "Protest—Past and Present." *American Scholar* 38: 535–46.

Capocci, M., and G. Corbellini. 2002. "Adriano Buzzati-Traverso and the Foundation of the International Laboratory of Genetics and Biophysics in Naples (1962–1969)." *Stud His Phil Biol Biomed Sci* 33: 489–513.

Carter, R. 1966. *Breakthrough: The Saga of Jonas Salk.* New York: Trident Press.

Cassidy, D.C. 2005. *J. Robert Oppenheimer and the American Century.* New York: Pi Press.

Cohn, M. 1967. "Reflections on a Discussion with Karl Popper: The Molecular Biology of Expectation." *Bulletin of the All-India Institute of Medical Sciences* 1: 8–16.

———. 1979. "In Memoriam to Jacques Monod." In *Origins of Molecular Biology: A Tribute to Jacques Monod,* ed. A. Lwoff and A. Ullmann, pp. 75–87. New York: Academic Press.

———. 1989. "The Way It Was." *Biochimica et Biophysica Acta* 1000: 109–12.

Cohn, M., and J. Monod. 1951. "Purification et propriétés de la beta-galactosidase (lactase) d'Escherichia coli." *Biochimica Biophysica Acta* 7: 153–74.

Coleman, P. 1989. *The Liberal Conspiracy: The Congress for Cultural Freedom and the Struggle for the Mind of Postwar Europe.* New York and London: Free Press/Collier Macmillan.

Crane, C.B. 1991. "The Pueblo Lands: San Diego Spanish Heritage." *Journal of San Diego History* 37.

Crichton, M. 1969. *The Andromeda Strain.* New York: Alfred A. Knopf.

Crick, F. 1979. "Sailing with Jacques." In *Origins of Molecular Biology: A Tribute to Jacques Monod,* ed. A. Lwoff and A. Ullmann, pp. 225–29. New York: Academic Press.

———. 1988. *What Mad Pursuit: A Personal View of Scientific Discovery.* New York: Basic Books.

Crotty, S. 2001. *Ahead of the Curve: David Baltimore's Life in Science.* Berkeley: University of California Press.

De Lattre de Tassigny, J. 1971. *Histoire de la Première Armée Française.* Paris: Presses de la Cité.

Debré, P. 1996. *Jacques Monod.* Paris: Flammarion.

Doolittle, R. 1974. "Antibody Diversity." *Science* 183: 190–91.

Dulbecco, R. 1952. "Production of Plaques in Monolayer Tissue Cultures by Single Particles of an Animal Virus." *Proceedings of the National Academy of Sciences USA* 38: 747–52.

———. 1990. *Aventurier du vivant.* Paris: Plon.

Dulbecco, R., and M. Vogt. 1954. "Plaque Formation and Isolation of Pure Lines with Poliomyelitis Viruses." *Journal of Experimental Medicine* 99: 167–82.

Dyson, F.J. 1981. *Disturbing the Universe.* New York: Harper & Row.

———. 2004. *Infinite in All Directions: Gifford Lectures given at Aberdeen, Scotland, April–November 1985.* New York: HarperCollins.

Dyson, G. 2002. *Project Orion: The True Story of the Atomic Spaceship.* New York: Henry Holt and Co.

Feld, B.T., and G.W. Szilard, eds. 1972. *Collected Works of Leo Szilard: Scientific Papers,* vol. 1. Cambridge, MA: MIT Press.

Fishbein, M. 1957. *Basil O' Connor: A Biographical Note,* vol. 5. New York: New York Academy of Sciences.

Flexner, A. 1960. *Abraham Flexner: An Autobiography.* New York: Simon and Schuster.

Fredrickson, D.S. 2001. *The Recombinant DNA Controversy: A Memoir.* Washington, DC: American Society for Microbiology.

Friedman, D.S. 1991. "Salk Institute for Biological Studies." In *Louis I. Kahn: In the Realm of Architecture,* ed. D.B. Brownlee and D.G. De Long, pp. 330–39. New York: Rizzoli International Publications.

Friedmann, T. 1971. "Prenatal Diagnosis of Genetic Disease." *Sci Am* 225: 34–42.

Friedmann, T., and R. Roblin. 1972. "Gene Therapy for Human Genetic Disease?" *Science* 175: 949–55.

Gaudillière, Jean-Paul. 2002. *Inventer la Biomédecine.* Paris: Editions La Découverte.

Gilot, F., and C. Lake. 1964. *Life with Picasso.* New York: McGraw-Hill Book Company.

Goldman, A., E. Schmalstieg, D. Freeman Jr., D. Goldman, and F. Schmalstieg Jr. 2003. "What Was the Cause of Franklin Delano Roosevelt's Paralytic Illness?" *Journal of Medical Biography* 11: 232–40.

Grémion, P. 1995. *Intelligence de l'Anticommunisme.* Paris: Fayard.

Gruss, P. 2012. "Driven by Basic Research." *Science* 336: 392.

Guillemin, R. 1978. "Pioneering in Neuroendocrinology 1952–1969." In *Pioneers in Neuroendocrinology II,* ed. J. Meites, B.T. Donovan, and S.M. McCann, pp. 220–39. New York: Plenum Press.

———. 1998. "Roger Guillemin." In *The History of Neuroscience in Autobiography,* vol. 2, ed. L.R. Squire, pp. 94–131. New York: Academic Press.

Hawkins, H.S., G.A. Greb, and G.W. Szilard, eds. 1987. *Toward a Livable World: Leo Szilard and the Crusade for Nuclear Arms Control.* Cambridge, MA: MIT Press.

Hayflick, L. 1966. "Cell Culture and the Aging Phenomenon." In *Topics in the Biology of Aging,* ed. P.L. Krohn, pp. 83–100. New York: John Wiley & Sons.

Herken, G. 2002. *Brotherhood of the Bomb: The Tangled Lives and Loyalties of Robert Oppenheimer, Ernest Lawrence, and Edward Teller.* New York: Henry Holt and Co.

Heuck, D. 1994. "Institute a Blueprint for a New Way of Thinking." *Pittsburgh Post-Gazette,* November 28.

Hilschmann, N., and L.C. Craig. 1965. "Amino Acid Sequence Studies with Bence-Jones Proteins." *Biochemistry* 53: 1403–9.

Hogness, D.S., M. Cohn, and J. Monod. 1955. "Studies on the Induced Synthesis of Beta-galactosidase in Escherichia coli: The Kinetics and Mechanism of Sulfur Incorporation." *Biochimica Biophysica Acta* 16: 99–116.

Hubel, D.A.H. 1963. "The Visual Cortex of the Brain." *Sci Am* 209: 54–62.

Hubel, D.H., and T.N. Wiesel. 1962. "Receptive Fields, Binocular Interaction and Functional Architecture in the Cat's Visual Cortex." *J Physiol* 160: 106–54.

Hunt, J. 1956. *Generations of Men.* Boston: Little, Brown.

———. 1968. *The Grey Horse Legacy.* New York: Knopf.

Hyman, S. 1975. *The Aspen Idea.* Norman: University of Oklahoma Press.

Jacob, F. 1988. *The Statue Within.* New York: Basic Books.

Jacobson, M.F. 1972. *Eater's Digest: The Consumer's Fact-book of Food Additives*. New York: Doubleday.

Johns, R.C. 1960. "Pact Provides 27-Acre Site for Institute." *San Diego Union*, May 25.

Judson, H.F. 1979. *The Eighth Day of Creation: Makers of the Revolution in Biology*. New York: Simon and Schuster.

Kay, L.E. 1993. *The Molecular Vision of Life: Caltech, The Rockefeller Foundation, and the Rise of the New Biology*. New York: Oxford University Press.

———. 2000. *Who Wrote the Book of Life? A History of the Genetic Code*. Stanford, CA: Stanford University Press.

Kelly, C.C. 2006. *Oppenheimer and the Manhattan Project: Insights into J. Robert Oppenheimer, "Father of the Atomic Bomb."* Hackensack, NJ: World Scientific Publishing.

Kevles, D.J. 1993. "Renato Dulbecco and the New Animal Virology: Medicine, Methods, and Molecules." *J Hist Biol* 26: 409–42.

Krohn, P.L., ed. 1966. *Topics in the Biology of Aging*. New York: John Wiley & Sons.

Lanouette, W. 1992. *Genius in the Shadows: A Biography of Leo Szilard: The Man Behind the Bomb*. With B.A. Silard. New York: C. Scribner's Sons.

Larson, H.J., and I. Ghinai. 2011. "Lessons from Polio Eradication." *Nature* 473: 446–47.

Larson, K. 2000. *Louis I. Kahn: Unbuilt Masterworks*. New York: Monacelli Press.

Lawrence, H.S. 1999. *Alwin Max Pappenheimer, Jr. 1908–1995*, vol. 77. Washington, DC: The National Academy Press.

Lennox, E.S. 1955. "Transduction of Linked Genetic Characters of the Host by Bacteriophage P1." *Virology* 1: 190–206.

Lennox, E.S., S.E. Luria, and S. Benzer. 1954. "On the Mechanism of Photoreactivation of Ultraviolet-Inactivated Bacteriophage." *Biochimica Biophysica Acta* 15: 471–74.

Leslie, T. 2005. *Louis I. Kahn: Building Art, Building Science*. New York: George Braziller.

Levi-Montalcini, R. 1988. *In Praise of Imperfection: My Life and Work*. New York: Basic Books.

Luria, S.E. 1984. *A Slot Machine, A Broken Test Tube (An Autobiography)*. New York: Harper & Row.

Luria, S.E., and M. Delbrück. 1943. "Mutations in Bacteria from Virus Sensitivity to Virus Resistance." *Genetics* 28: 491–511.

Lwoff, A., and A. Ullmann, eds. 1978. *Selected Papers in Molecular Biology by Jacques Monod*. New York: Academic Press.

———. 1979. *Origins of Molecular Biology: A Tribute to Jacques Monod*. New York: Academic Press.

Manley, J.H. 1974. "All in Our Time: Assembling the Wartime Labs." *Bulletin of the Atomic Scientists*, 42–48.

Mason, M., and W. Weaver. 1929. *The Electromagnetic Field*. Chicago: University of Chicago Press.

Meitner, L., and O. Frisch. 1939. "Disintegration of Uranium by Neutrons: A New Type of Nuclear Reaction." *Nature* 143: 239–40.

Morange, Michel. 1998. *A History of Molecular Biology.* Cambridge, MA: Harvard University Press.

Morin, E. 1970. *Journal de Californie.* Paris: Editions du Seuil.

———. 1973. *Le Paradigme Perdu: La Nature Humaine.* Paris: Editions du Seuil.

Morin, E., and M. Piattelli-Palmarini, eds. 1974. *L'Unité de l'Homme.* Paris: Editions du Seuil.

National Foundation News. 1972. Vol. 29: 1.

Nelson, B. 1968. "Edward Harold Litchfield: An Administrative Career Cut Short." *Science* 159: 1212–14.

Offit, P. A. 2005. *The Cutter Incident: How America's First Polio Vaccine Led to the Growing Vaccine Crisis.* New Haven, CT: Yale University Press.

Olby, R. 2009. *Francis Crick: Hunter of Life's Secrets.* New York: Cold Spring Harbor Laboratory Press.

Orgel, L. E. 1963. "The Maintenance of the Accuracy of Protein Synthesis and Its Relevance to Aging." *Proceedings of the National Academy of Sciences USA* 49: 517–21.

Orrick, P. 1996. "Stranded in La Jolla." *The Reader* (San Diego) 25: 10–30.

Oshinsky, D. M. 2005. *Polio: An American Story.* New York: Oxford University Press.

Pappenheimer, A. M., Jr. 1937. "Diphteria Toxin. I. Isolation and Characterization of a Toxic Protein from Corynebacterium Diphtheriae Filtrates." *J Biol Chem* 120: 543–53.

Paul, J. R. 1971. *A History of Poliomyelitis.* New Haven, CT: Yale University Press.

———. 1974. "Thomas Francis, Jr., 1900–1969." In *Biographical Memoir.* Washington, DC: National Academy of Sciences.

Paull, B. I. 1986. *A Century of Medical Excellence.* Pittsburgh, PA: University of Pittsburgh Medical Alumni Association.

Piattelli-Palmarini, M., ed. 1979. *Théories du langage, théories de l'Apprentissage.* Paris: Editions du Seuil.

Platt, J. R. 1969. "What We Must Do." *Science* 166: 1115–21.

Podolsky, S. H., and A. I. Tauber. 1997. *The Generation of Diversity.* Cambridge, MA: Harvard University Press.

"Polio Fund Owes $ 2,000,000; Asks Hospitals to Defer Debts." 1960. *New York Times,* July 1.

Popper, K. R. 1973. *La Logique de la Découverte Scientifique.* Paris: Payot.

Porter, R. R. 1986. "Antibody Structure and the Antibody Workshop 1958–1965." *Perspect Biol Med* 29: S161–65.

Potter, M. 2007. "The Early History of Plasma Cell Tumors in Mice, 1954–1965." In *Adv Cancer Res* 98: 17–51.

Raitt, H., and B. Moulton. 1967. *Scripps Institution of Oceanography; First Fifty Years.* Los Angeles: Ward Ritchie Press.

Rees, M. 1980. "The Mathematical Sciences and World War II." *American Mathematical Monthly* 87: 607–21.

Regis, E. 1987. *Who Got Einstein's Office? Eccentricity and Genius at the Institute for Advanced Study.* Cambridge, MA: Perseus Publishing.

"Research Institute to Be Established in California." 1960. *Science* 131: 1088.

Rhodes, R. 1986. *The Making of the Atomic Bomb.* New York: Simon & Schuster.

Roberts, L. 2010. "Polio Outbreak Breaks the Rules." *Science* 330: 1730–31.

Salk, J. E. 1955. "Vaccines for Poliomyelitis." *Sci Am* 194: 42–44.

Sanchez, R. A., and L. E. Orgel. 1970. "Studies in Prebiotic Synthesis. V. Synthesis and Photoanomerization of Pyrimidine Nucleosides." *J Mol Biol* 47: 531–43.

Saunders, F. S. 1999. *Who Paid the Piper? The CIA and the Cultural Cold War.* London: Granta Books.

Scherer, G., and M. Fletcher. 2008. *J. Robert Oppenheimer: The Brain Behind the Bomb.* Berkeley Heights, NJ: MyReportLinks.com.

Schrödinger, E. 1992. Autobiographical sketches in *What is Life?*, pp. 167–84. Cambridge: Cambridge University Press.

Scudellari, M. 2010. "Nice Shot." *The Scientist* 24: 32–37.

Serber, R. 1998. *Peace and War: Reminiscences of a Life on the Frontier of Science.* New York: Columbia University Press.

Shannon, C. E., and W. Weaver. 1949. *The Mathematical Theory of Communication.* Urbana: University of Illinois Press.

Sher, A. 1981. "Training and Career Opportunities in Parasitology: Entering the Field from Another Discipline." In *The Current Status and Future of Parasitology,* ed. K. S. Warren and E. F. Pucell, pp. 103–8. New York: Josiah Macy Jr. Foundation.

Shils, E. 1990. "Remembering the Congress for Cultural Freedom." *Encounter* 75: 53–65.

Shor, E. N. 1978. *Scripps Institution of Oceanography: Probing the Oceans, 1936 to 1976.* San Diego, CA: Tofua Press.

Sills, D. L. 1980. *The Volunteers: Means and Ends in a National Organization.* New York: Arno Press.

Sime, R. L. 1996. *Lisa Meitner: A Life in Physics.* Berkeley: University of California Press.

Sloan, P. R., and B. Fogel, eds. 2011. *Creating a Physical Biology: The Three-Man Paper and Early Molecular Biology.* Chicago: University of Chicago Press.

Smith, J. S. 1990. *Patenting the Sun: Polio and the Salk Vaccine.* New York: W. Morrow.

Snow, C. P. 1959. *The Two Cultures.* London: Cambridge University Press.

Solow, H. 1958. "The All-Purpose Executive." *Fortune,* December, 122–24, 190, 193.

Soulier, J.-P. 1997. *Jacques Monod: Le Choix de l'Objectivité.* Paris: Editions Frison-Roche.

Strickland, R. 2005. "Nothing That Is Human Is Alien to Me: Neoliberalism and the End of Bildung." *RiLUnE* 1: 29–40.

Sullerot, E., ed. 1978. *Le Fait Féminin.* Paris: Fayard.

Summers, W. C. 1999. *Félix d'Herelle and the Origins of Molecular Biology*. New Haven, CT: Yale University Press.

Szilard, L. 1959. "On the Nature of the Aging Process." *Proceedings of the National Academy of Sciences USA* 45: 30–45.

———. 1976. *Die Stimme der Delphine*. Munich: Wilhelm Heyne Verlag.

Takaro, T. 2004. "The Man in the Middle." *Dartmouth Medicine* 29.

Teller, E. 2001. *Memoirs : A Twentieth-Century Journey in Science and Politics*. With J. L. Shoolery. Cambridge, MA: Perseus Publishing.

Thorpe, C., and S. Shapin. 2000. "Who Was J. Robert Oppenheimer? Charisma and Complex Organization." *Social Studies of Science* 30: 545–90.

Timoféeff-Ressovsky, N. W., K. G. Zimmer, and M. Delbrück. 1935. "Über die Natur der Genmutation und der Genstruktur." *Nachr Ges Wiss Gottingen, math phys* 6: 189–245.

Watson, J. D. 2007. *Avoid Boring People*. New York: Random House.

Watson, J. D., and F. Crick. 1953a. "Genetical Implications of the Structure of Deoxyribonucleic Acid." *Nature* 171: 964–67.

———. 1953b. "Molecular Structure of Nucleic Acids: A Structure for Deoxyribose Nucleic Acid." *Nature* 171: 737–38.

Weart, S. R., and G. W. Szilard, eds. 1978. *Leo Szilard: His Version of the Facts—Selected Recollections and Correspondence*. Cambridge, MA: MIT Press.

Weaver, W. 1964. *Alice in Many Tongues*. Madison: University of Wisconsin Press.

———. 1970a. "Molecular Biology: Origin of the Term." *Science* 170: 581–82.

———. 1970b. *Scene of Change: A Lifetime in American Science*. New York: Scribner.

Weir, J. M. 1969. "The Unconquered Plague." *Rockefeller Foundation Quarterly* 2: 4–23.

Weiss, P. A. 1971. *Within the Gates of Science and Beyond*. New York: Hafner Publishing Company.

Wiseman, C. 2007. *Louis I. Kahn, Beyond Time and Style: A Life in Architecture*. New York: W. W. Norton & Co.

Woodward, R. B., and W. E. Doering. 1944. "The Total Synthesis of Quinine." *Journal of the American Chemical Society* 66: 849.

Index